Infrared Spectroscopy

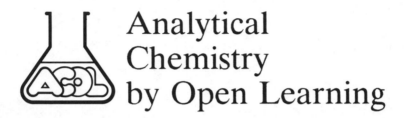

Analytical Chemistry by Open Learning

Titles in Series:

Infrared Spectroscopy

Analytical Chemistry by Open Learning

Authors:
W. O. GEORGE
P. S. MCINTYRE
Both at Polytechnic of Wales, Pontypridd

Editor:
DAVID J. MOWTHORPE

on behalf of ACOL

Published on behalf of ACOL, Thames Polytechnic,
London
by
JOHN WILEY & SONS
Chichester · New York · Brisbane · Toronto · Singapore

Library of Congress Cataloging in Publication Data:

George, W. O. (William O.)
  Infrared spectroscopy.

  (Analytical chemistry by open learning)
  Bibliography: p.
  1. Infrared spectroscopy—Programmed instruction.
  2. Chemistry, Analytic—Programmed instruction.
  I. McIntyre, P. II. Mowthorpe, David J. III. ACOL
  (Firm : London, England) IV. Title. V. Series.
  QD96.I5G46 1987   543'.08583   87-2110
  ISBN 0 471 91382 0
  ISBN 0 471 91383 9 (pbk.)

British Library Cataloguing in Publication Data:

George, W.O.
  Infrared spectroscopy.—(Analytical
  chemistry).
  1. Infrared spectroscopy
  I. Title II. McIntyre, P.
  III. Mowthorpe, David J. IV. ACOL
  V. Series
  535.8'42   QC457

  ISBN 0 471 91382 0
  ISBN 0 471 91383 9 Pbk

Printed and bound in Great Britain

Analytical Chemistry

This series of texts is a result of an initiative by the Committee of Heads of Polytechnic Chemistry Departments in the United Kingdom. A project team based at Thames Polytechnic using funds available from the Manpower Services Commission 'Open Tech' Project has organised and managed the development of the material suitable for use by 'Distance Learners'. The contents of the various units have been identified, planned and written almost exclusively by groups of polytechnic staff, who are both expert in the subject area and are currently teaching in analytical chemistry.

The texts are for those interested in the basics of analytical chemistry and instrumental techniques who wish to study in a more flexible way than traditional institute attendance or to augment such attendance. A series of these units may be used by those undertaking courses leading to BTEC (levels IV and V), Royal Society of Chemistry (Certificates of Applied Chemistry) or other qualifications. The level is thus that of Senior Technician.

It is emphasised however that whilst the theoretical aspects of analytical chemistry can be studied in this way there is no substitute for the laboratory to learn the associated practical skills. In the U.K. there are nominated Polytechnics, Colleges and other Institutions who offer tutorial and practical support to achieve the practical objectives identified within each text. It is expected that many institutions worldwide will also provide such support.

The project will continue at Thames Polytechnic to support these 'Open Learning Texts', to continually refresh and update the material and to extend its coverage.

Further information about nominated support centres, the material or open learning techniques may be obtained from the project office at Thames Polytechnic, ACOL, Wellington St., Woolwich, London, SE18 6PF.

How to Use an Open Learning Text

Open learning texts are designed as a convenient and flexible way of studying for people who, for a variety of reasons cannot use conventional education courses. You will learn from this text the principles of one subject in Analytical Chemistry, but only by putting this knowledge into practice, under professional supervision, will you gain a full understanding of the analytical techniques described.

To achieve the full benefit from an open learning text you need to plan your place and time of study.

- Find the most suitable place to study where you can work without disturbance.

- If you have a tutor supervising your study discuss with him, or her, the date by which you should have completed this text.

- Some people study perfectly well in irregular bursts, however most students find that setting aside a certain number of hours each day is the most satisfactory method. It is for you to decide which pattern of study suits you best.

- If you decide to study for several hours at once, take short breaks of five or ten minutes every half hour or so. You will find that this method maintains a higher overall level of concentration.

Before you begin a detailed reading of the text, familiarise yourself with the general layout of the material. Have a look at the course contents list at the front of the book and flip through the pages to get a general impression of the way the subject is dealt with. You will find that there is space on the pages to make comments alongside the text as you study—your own notes for highlighting points that you feel are particularly important. Indicate in the margin the points you would like to discuss further with a tutor or fellow student. When you come to revise, these personal study notes will be very useful.

∏   When you find a paragraph in the text marked with a symbol such as is shown here, this is where you get involved. At this point you are directed to do things: draw graphs, answer questions, perform calculations, etc. Do make an attempt at these activities. If necessary cover the succeeding response with a piece of paper until you are ready to read on. This is an opportunity for you to learn by participating in the subject and although the text continues by discussing your response, there is no better way to learn than by working things out for yourself.

We have introduced self assessment questions (SAQ) at appropriate places in the text. These SAQs provide for you a way of finding out if you understand what you have just been studying. There is space on the page for your answer and for any comments you want to add after reading the author's response. You will find the author's response to each SAQ at the end of the text. Compare what you have written with the response provided and read the discussion and advice.

At intervals in the text you will find a Summary and List of Objectives. The Summary will emphasise the important points covered by the material you have just read and the Objectives will give you a checklist of tasks you should then be able to achieve.

You can revise the Unit, perhaps for a formal examination, by re-reading the Summary and the Objectives, and by working through some of the SAQs. This should quickly alert you to areas of the text that need further study.

At the end of the book you will find for reference lists of commonly used scientific symbols and values, units of measurement and also a periodic table.

Contents

Study Guide

The purpose of this Unit is to equip you, the learner, with the knowledge to understand infrared spectroscopy and apply it to the solution of analytical and structural problems.

We use the title 'infrared spectroscopy' to define a specific region of the electromagnetic spectrum. This region is much used in industry and elsewhere. The results are unusual because frequently they can apparently be interpreted in simple classical empirical terms; this means some care should be taken to grasp the connections with related spectroscopic methods. You may already have a working knowledge of some aspects of the subject. No one person can have a working knowledge of all aspects of infrared spectroscopy and the vast majority of us are somewhere in the middle ground! To cover the needs of learners who know either very little, or a fair amount of particular aspects, the Unit is divided into seven fairly self-contained Parts. A glance at the learning objectives and the SAQ's associated with the various Parts will tell you whether you are in either new or familiar territory.

Little pre-knowledge is required other than the concepts of physics to about BTEC certificate or diploma in a relevant subject, or 'A' level; and about the same level of the vocabulary of organic chemistry where most of the basic applications occur. Many very good spectroscopists have no formal chemical education.

The first Part is an introduction which places infrared spectroscopy in a general context and provides a basic foundation of theory. If you lack a philosophical approach and simply want to use the method to study organic problems quickly and empirically you may want to skip over some of the first Part.

The second Part concerns instrumentation, dispersive, Fourier transform methods and computer interfacing. It links closely to the third Part which concerns sample handling techniques and the various accessories.

The fourth Part deals with spectrum interpretation from the point of view of group frequencies and links to the important special case of hydrogen bonding in the fifth Part.

The sixth Part looks at quantitative analysis of various kinds and leads into a general consideration of structure determination in the seventh Part.

Practical Objectives

It would be highly desirable to have between 6 and 12 hours of practical time in each of the following two areas.

1. Instrumentation and Sample Handling

This would involve using a range of spectrometers and sample accessories to record spectra of routine and more difficult samples. By the end of the course the learner should recognise and possess the skills necessary to record good quality spectra.

2. Spectra Handling and Interpretation

Here the challenge is to interpret spectra and to store and retrieve spectra from libraries and computer data collections. The learner should quickly be able to handle and interpret large numbers of spectra.

Bibliography

The following books provide a general introduction to Molecular Spectroscopy:

1. E F H Brittain, W O George and C H J Wells. *Introduction to Molecular Spectroscopy – Theory and Experiment*, Academic Press, 1975.

2. R C J Osland, *Principles and Practices of Infrared Spectroscopy*, Philips, 1985.

The following books are more advanced treatments of the fundamental theory of vibrational spectroscopy in relation to infrared and Raman.

3. J R Durig (Ed.), *Chemical, Biological and Industrial Applications of Infrared Spectroscopy*, Wiley 1985.

4. J M Hollas, *High Resolution Spectroscopy*, Butterworths, 1982.

The following books relate to Fourier-Transform Infrared (Ft-ir)

5. P R Griffiths and J A de Haseth, *Fourier Transform Infrared Spectroscopy, Chemical Analysis Series, Vol. 83*, John Wiley and Sons, 1986.

Finally there are various audio/visual/computer aided learning aids including the following:

6. D A Kealey, *Introduction to Infrared Spectroscopy*, Wiley 1986, Computer aided interactive program on Discs to run on a BBC Microcomputer.

Acknowledgements

Figures 1.1a, 1.1b, 1.1c, 1.1e, 1.2a, 1.2d, 1.3b, 1.3d, 1.3f, 1.5b, 1.5f and 2.1a are redrawn from E. F. M. Brittan, W. O. George and C. H. J. Wells, *Introduction to Molecular Spectroscopy*, Academic Press, 1972. Permission has been requested.

Figure 1.2b is redrawn from D. A. C. Comptom, W. O. George, Q. J. Haines, *Education in Chemistry*, **19**, *13*, 1976. Permission has been requested from the Royal Society of Chemistry.

Figure 1.4a is redrawn from J. M. Hollas, *High Resolution Spectroscopy*, Butterworths, 1982, with permission of Butterworth and Company.

Figures 1.5c, 1.5d and 1.5e are redrawn from G. Herzberg, *Molecular Spectra of Polyatomic Molecules*, 1145, Van Nostrand, 1954. Permission has been requested.

Figures 2.1c, 2.1d, 2.2a, 2.2b and 2.2d are redrawn from A. J. Barnes, W. J. Orville-Thomas (Ed), *Vibrational Spectroscopy – Modern Trends*, Elsevier, 1977 with permission.

Figure 2.2c is redrawn from P. R. Griffiths, *Chemical Infrared Fourier Transform Spectroscopy*, Wiley 1975. © John Wiley & Sons Inc. Reprinted by permission of John Wiley & Sons Inc.

Figures 2.2e, 2.2f and 2.2g are redrawn from a Perkin Elmer sales pamphlet; Figures 6.4e and 6.4f are redrawn from the Perkin Elmer *Application Note on Derivative Spectroscopy*. © Perkin–Elmer, 1984. All reproduced by permission of Perkin–Elmer Ltd.

Figures 3.4a, 3.5b and 3.5c are redrawn from R. R. Hill and D. A. E. Rendell, *The Interpretation of Infrared Spectra: a programmed introduction*, Heyden, 1975. © Heyden & Son Ltd. Reprinted by permission of John Wiley & Sons Ltd.

Figure 3.4b is reproduced from a Pye Unicam instrument manual by permission of Pye Unicam, Cambridge.

1. Introduction to Infrared Spectroscopy

1.1. GENERAL INTRODUCTION TO THE ELECTROMAGNETIC SPECTRUM

To some the electromagnetic spectrum is a formidable subject so why not start with something familiar? The spectrum of visible light has been long observed in nature in the form of the Rainbow. An early reference is in Genesis, ch 9, verse 13.

'I do set my bow in the cloud, and it shall be for a token of a covenant between me and the earth'.

Since then much literature, art and music has been inspired by this phenomenon. An impressive scene carrying the previous quotation is set in stained glass in Worcester Cathedral to commemorate Sir Edward Elgar.

Many scientists speculated on the origin of the rainbow until Newton in 1666 bridged the cultures of science and art most decisively by demonstrating that radiation from the sun could be split into the component colours of the rainbow using a glass prism as shown in Fig. 1.1a.

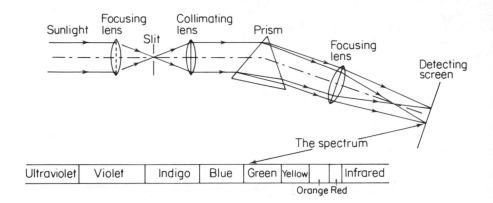

Fig. 1.1a.   *The formation of the visible spectrum*

You may find it interesting or helpful to demonstrate this effect. One way is by taking a slide projector and mounting an opaque square containing a narrow vertical slit in the slide position in front of which a prism can be positioned to project the visible spectrum on the screen. In Fig. 1.1a lenses are shown which collimate and focus the beam. These can improve the quality of the spectral image but are not essential. You may also have seen similar spectral images from bevelled mirrors or other suitably refracting surfaces.

We have described, and you may have been able to demonstrate, the following basic components of a dispersive spectrometer viz:

(*a*) A source of radiation consisting of a filament of a bulb with the radiation focussed on a slit,

(*b*) A dispersion element consisting of a prism,

(*c*) A detection system consisting of a screen.

It would be unsatisfactory to leave the comparison between the natural rainbow and the visible spectrum without asking if there are any observable differences between the two phenomena and if so how are they explained?

When looking at a strong natural rainbow can you observe the following features?

(*a*) The primary rainbow and colour sequence
(*b*) The secondary rainbow and colour sequence
(*c*) The supernumerary arcs below the primary rainbow
(*d*) The dark space between the two bows
(*e*) The light spaces above and below the two bows.

A full understanding of these observations is not necessary for the purpose of this unit of work but if you have an enquiring mind I can recommend an article in 'Scientific American' 236, 116 (1977) which shows that although the main features of the rainbow have been understood for a long time, some of the details have only recently been explained.

1.1.1. The Electromagnetic Spectrum

We have observed a continuous visible spectrum obtained from the sun and from a projector bulb which contains an incandescent tungsten filament. This radiation is, by definition, visible to the human eye, it is also continuous or polychromatic. Other detection systems reveal radiation beyond the visible regions of the spectrum which are classified in an arbitrary manner as γ-ray, x-ray, ultraviolet, visible, infrared, microwave and radiowave. These regions are shown in Fig. 1.1b together with the processes involved in the interaction of the radiation in these regions with matter.

The visible region is a relatively small portion of the total. Other regions are described by their historical or technical origins. The radiations in these regions interact with matter in many different ways by processes which are summarised in Fig. 1.1b but are considered in detail elsewhere. The electromagnetic spectrum (the e.m. spectrum for brevity) and the varied interactions between these radiations and many forms of matter can either be considered in terms of classical or quantum theories.

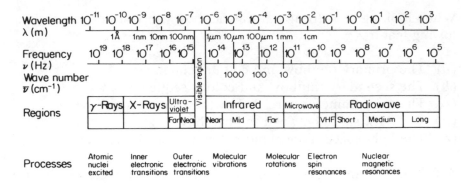

Fig. 1.1b.   *The electromagnetic spectrum*

1.1.2.   The Classical Wave Theory of Electromagnetic Radiation

The nature of the various radiations shown in Fig. 1.1b have been interpreted by Maxwell's classical theory of electro- and magneto-dynamics. Hence the term 'electromagnetic radiation'; thus radiation is considered as two mutually perpendicular electric and magnetic fields, oscillating in single planes at right angles to each other and in phase and being propagated as a sine wave as shown in Fig. 1.1c. The electric and magnetic vectors are represented by E and B respectively.

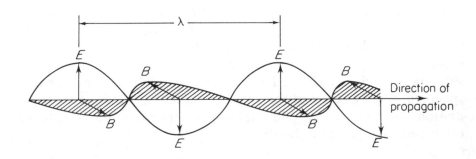

Fig. 1.1c.   *The electromagnetic wave*

The classical model of e.m. radiation may be understood at various levels and the depth of understanding needed depends on what you may wish to use the information for. Initially we need to have a picture of the kind shown in Fig. 1.1c.

A remarkable discovery about electromagnetic radiation was that the velocity of propagation in a vacuum was constant for all regions of the spectrum. This is known as the velocity of light and has the value

$$c = 2.997925 \times 10^8 \text{ m s}^{-1}$$

In any homogeneous medium of refractive index, n, the velocity is c/n. If you can visualize one complete wave travelling a fixed distance each cycle you should, with a little imagination, be able to see that the velocity of this wave is the product of the wavelength, λ (or distance between adjacent peaks), and the frequency, ν (or number of cycles per second). It follows that

$$c = n\lambda\nu \tag{1.1}$$

The refractive index, n, is the ratio of the velocity of e.m. radiation in vacuo to that in the homogeneous medium in which the radiation is propagated. The value of n is close to 1 in air and 1.6 in glass when measured in terms of polychromatic or 'white radiation'.

The value of n at particular wavelengths of monochromatic radiation may be different for the same medium which leads to the dispersion of polychromatic radiation into monochromatic components as with a glass prism in Fig. 1.1a or by a shower of raindrops (water has a slightly smaller refractive index than glass) in a rainbow.

SAQ 1.1a   Give two examples of sources of continuous or polychromatic radiation.

SAQ 1.1a

SAQ 1.1b   Assuming n = 1.68 and 1.64 for violet and red rays respectively sketch the path of sunlight through

(i)   a glass prism

(ii)   a raindrop.

SAQ 1.1c

Comment on the sequence of colours in

(*i*) the visible spectrum

(*ii*) the primary rainbow

(*iii*) the secondary rainbow.

Your success or otherwise in achieving reasonable answers depends on your knowledge of basic optics such as Snell's law. If this is rusty or non-existent you may like to consult a basic physics text.

The presentation of spectral regions may be in terms of wavelength as metres or sub-multiples of a metre.

$1 \text{ Å} = 10^{-10} \text{ m}$, $1 \text{ nm} = 10^{-9} \text{ m}$, $1 \text{ } \mu\text{m} = 10^{-6}\text{m}$, $1 \text{ mm} = 10^{-3} \text{ m}$, $1 \text{ cm} = 10^{-2} \text{ m}$.

From the wavelength of any radiation the frequency may be calculated in hertz (cycles per second) using the equation

$$\nu = \frac{c}{\lambda} \tag{1.2}$$

A third unit which is used particularly in the optical regions of the spectrum (ultraviolet, visible, infrared) is the wavenumber in cm^{-1}, which is the number of waves in a length of one centimetre and is given by the relationship

$$\nu = \frac{1}{\lambda} = \frac{\nu}{c} \tag{1.3}$$

This unit is the reciprocal of wavelength but unlike wavelength has the advantage of being linear with energy. We should attempt to use terms like '–the wavenumber value of the carbonyl band is 1738 cm^{-1}–' rather than '–the frequency value of the carbonyl band is 1738 cm^{-1}–'

SAQ 1.1d   The following diagram shows regions of the e.m. spectrum with arbitrary intervals measured in metres.

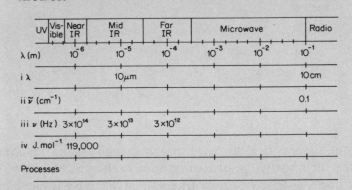

Fig. 1.1d.   *The electromagnetic spectrum*

→

SAQ 1.1d (cont.)

Complete the diagram as follows:

(*i*) enter values in suitable sub-multiples of metres in line i

(*ii*) enter values in wavenumber units in line ii

(*iii*) enter values in frequency units in line iii.

SAQ 1.1e

What changes in molecular energies are associated with absorption of radiation in the regions

(*i*) microwave/far infrared

(*ii*) mid infrared/far infrared?

SAQ 1.1e

Indicate the range of these processes in the last line of Fig. 1.1d. Before proceeding check your entry with the response to this SAQ given at the end of the Unit. You need to refer to this figure in a later section.

1.1.3. Quantum Theory of Electromagnetic Radiation

During the 19th century a number of experimental observations were made which were not consistent with the classical view that matter could interact with energy in a continuous form. For example Balmer noted that the spectrum of atomic hydrogen consisted of discrete lines for which the frequencies were interrelated as simple functions of integers. The spectrum of hydrogen was found to consist of discrete monochromatic lines. In effect Balmer played a simple numbers game the prize for which was the discovery of the principal quantum numbers. Subsequent work by Einstein, Planck and Bohr indicated that in many ways e.m. radiation could be regarded as a stream of particles or quanta for which the energy, ϵ, is given by the Bohr equation

$$\epsilon = h\nu \qquad (1.4)$$

where h is Planck's constant

$h = 6.6262 \times 10^{-34} \text{ J s}$

and ν is equivalent to the classical frequency.

It follows that high frequency (or short wavelength) radiation is of high energy. This is consistent with the effect of radiation beyond the violet (ultraviolet and of higher frequency) on organic tissue which frequently leads to photochemical change and other radiation damage whereas for radiation beyond the red (infrared and of lower frequency) no permanent chemical or physical change normally occurs. The energy of the quantum per atom or molecule can be calculated using Eq. 1.4. The energy per mole can be determined by multiplying the energy per molecule by the Avogadro number, N_A.

$N_A = 6.02205 \times 10^{23} \text{ mol}^{-1}$

SAQ 1.1f   Give two examples of sources of monochromatic radiation.

SAQ 1.1g | Complete line iv of Fig. 1.1d by calculating the energy in joules per mole for each of the wavelengths listed in the first line of that figure.

As you meet different experiments and examples of spectroscopic phenomena and interactions which may be interpreted by classical or quantum theories you may reflect on the questions:

(*a*) Are these two theories independent?

(*b*) Is one of them more satisfactory in some circumstances?

(*c*) Is one more satisfactory in all circumstances?

(*d*) Does one provide a more conceptually satisfying physical model than the other?

(*e*) Are they approximate models which only partly explain observations?

(f) Are we using a model which is a composite of a classical model and a quantum model?

These questions emerge in the consideration of electromagnetic radiation and throughout any study of spectroscopy. The answers are both subjective and dependent on the particular example but already you will have noticed that the wave and particle models are:

(a) Mutually dependent – both need ν.

(b) Each is more satisfactory in certain circumstances.

(c) Neither appears to be more satisfactory in all circumstances

(d) Probably the wave theory is more conceptually satisfying.

(e) It is too soon to answer this question since properties such as polarisation, coherence and diffration need to be considered.

(f) Already we can see the common feature of frequency in both models. When we come to consider potential energies of molecules we will note the composite classical and quantum nature of our model. At this stage it is apparent you should bear in mind these questions throughout this unit although you may notice that experienced spectroscopists reach for a model which best serves their purpose without too much regard to its classical or quantum parentage.

1.1.4. Experimental Arrangements for Obtaining Spectra

A schematic form of the arrangements for obtaining spectra is shown in Fig. 1.1e.

The top arrangement shows how an emission spectrum may be obtained. The sample is excited in some way and emits radiation. Some of the radiation is collected by a lens and focussed on the spectrometer. The spectrometer sorts out the e.m. radiation allowing an emission spectrum to be displayed. This spectrum is effectively a

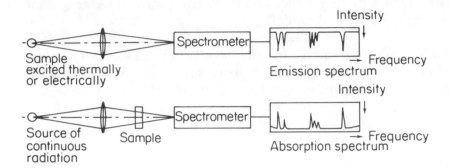

Fig. 1.1e.   *Experimental arrangements for obtaining emission and absorption spectra*

graph which displays the intensity of radiation emitted by the sample as a function of frequency.

The lower arrangement shows how an absorption spectrum may be obtained. The sample is no longer the source of radiation, but is now placed so that it can absorb radiation from an independent source, prior to analysis and display by the spectrometer. This spectrum is effectively a graph which displays the intensity of radiation absorbed by the sample as a function of frequency. You will see that the sample absorbs strongly at some frequencies, weakly at others and not at all in some regions of the spectrum.

Matter in molecular form is often decomposed by the conditions used to excite emission spectra and for this reason molecular spectroscopy is mostly concerned with absorption spectra. By contrast matter in atomic form can be subjected to electrical or thermal energy leading to emission of radiation the spectrum of which may be recorded. If you have seen an atomic arc, spark or flame source in action you will see why typical molecules are readily decomposed. Nevertheless under suitably mild conditions it is feasible to get molecular emission spectra. When atomic or molecular systems can yield spectra both in emission and absorption these spectra are typically coincident as shown in Fig. 1.1e. This correspondence indicates the sample undergoes the same energy change during the interaction. This implies that energy corresponding to a particular process may be either emitted or absorbed by a sample according to the experimental arrangement.

SAQ 1.1h

In the experimental arrangement for obtaining an emission spectrum (Fig. 1.1e) the sample itself emits radiation in 6 spectral regions. This is therefore, a line source since a series of frequencies (lines) are emitted with many regions of the spectrum devoid of radiation.

In the case of the arrangement for obtaining an absorption spectrum we observed the following.

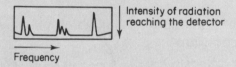

Intensity of radiation reaching the detector

Frequency

If we remove the sample no absorption of radiation takes place, and we are left with the emission spectrum of the source (a continuous source).

Draw the emission spectrum of this source; it is called the background.

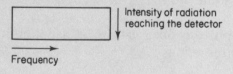

Intensity of radiation reaching the detector

Frequency

1.2. THE CLASSICAL THEORY OF VIBRATIONAL AND
     ROTATIONAL SPECTROSCOPY

The aim of the second Section of the first Part of the Unit is to understand the interactions of infrared and microwave radiations with matter in terms of the changes in molecular dipoles associated with vibrations and rotations described by simple classical potential functions.

So far, in Section 1.1, we have limited our attention to a broad look at the e.m. spectrum. We described e.m. radiation both in classical terms in which aspects were quantified by the relationship

$$c = n\lambda\nu \qquad (1.1)$$

and also in quantum terms in which other aspects were quantified by the relationship

$$\epsilon = h\nu \qquad (1.4)$$

We also posed some questions on the relationship between quantum and classical theory. The emphasis here is on 'we' rather than 'you' or 'I' because these are philosophical questions concerning physical models for observed phenomena. Another philosophical question concerns the nature of e.m. radiation in isolation or in abstract. Radiation only manifests itself when it interacts with some form of matter by some atomic or molecular process.

For example the observation of a coloured object implies three kinds of interactions.

(*a*) An interaction at the source of light leading to emission of visible radiation.

(*b*) An interaction at the reflecting surface of the object leading to absorption of some of the incident radiation prior to reflection.

(*c*) An interaction of the reflected light at the eye of the observer leading to some kind of signal to the brain.

These are examples of processes involving changes in electronic structure, within the respective materials of the source, object and human eye.

Let us now consider spectral features arising from molecular vibrations and molecular rotations. These are taken together because there is an obvious mechanical connection with large observable bodies such as we find in vehicles or fairgrounds. If we consider vibrations in terms of a fairground swing, or rotations in terms of a merry-go-round, these bodies are typically 10 m in size. Do you believe there is a contradiction in applying ideas derived from such large systems to molecules as small as 10 nm in size? Since a molecule is the microscopic concept for the ultimate unit of a compound can it be described in macroscopic terms? The answer probably is that this is where we have to start because it is the way our minds work. Maybe we will have to struggle to make our minds work in different ways before we finish!

The realization of the association between interaction of radiation in a particular region of the spectrum with molecular vibrations and rotations, as the processes responsible for this interaction emerged during the early decades of this century more as a collective view than as a brilliant insight by a single individual. It could be a fascinating exercise in scholarship to trace the evolution of this understanding but this is not the aim of our present section of work. It is more profitable to consider the extent to which these macroscopic models fit our microscopic problems.

Let us look further at Fig. 1.1d in which the vibration and rotation regions overlap. In classical terms, by which we mean applying the laws relating to macroscopic models, molecules may have high or low vibrational energy. This leads to interaction with e.m. radiation of high or low frequency. These interactions are found between 4000 cm^{-1} and 40 cm^{-1} in the spectrum. Similarly molecules may have high or low rotational energy leading to interactions between about 200 cm^{-1} and 0.1 cm^{-1}.

SAQ 1.2a   From Fig. 1.1d complete the following

(*i*)   The infrared/microwave interface is
                          m

(*ii*) The infrared/microwave interface is
                          cm^{-1}

(*iii*) The region indicated for molecular vibra-
        tions is from   to   m, or
        from   to cm^{-1}

(*iv*) The region indicated for molecular rota-
        tions is from   to   m, or
        from   to   Hz.

1.2.1. Classical Theory of Vibrational Interactions

Interactions between the electrical components of the e.m. radiation and electrical dipolar motions within molecules are the basis of classical theory. In the case of vibrational spectroscopy the dipolar motions are taken to be molecular vibrations leading to absorption in the infrared region. It is informative to consider how far spectral observations in the infrared can be explained in terms of a classical model.

In the case of a diatomic molecule it is observed that if it is a heteronuclear molecule (A-B) there is a strong absorption band centred at a particular wavenumber which is characteristic of that molecule. By contrast homonuclear molecules (A-A) do not absorb infrared radiation. For example hydrogen chloride (HCl) has one strong absorption band in the infrared centred at 2886 cm^{-1} and carbon monoxide has one strong absorption band centred at 2143 cm^{-1}. Diatomics such as hydrogen (H_2) and chlorine (Cl_2) are completely transparent to infrared radiation. If you happen to have seen an infrared spectrum of HCl or CO you may be puzzled about the shape or structure of the respective band which will also depend on the resolution qualities of the spectrometer used.

Consider a classical model of a diatomic molecule as two masses m_1 and m_2 joined together by a bond which is considered to behave like a spring which maintains the two masses at some equilibrium separation. The stiffness of the spring can be characterized by a constant termed the force constant, f. Let us assume the model obeys Hooke's Law which states that if a spring is displaced by a small distance, x, there will be an opposing restoring force, F, which will be proportional to the magnitude of the displacement. (Note the assumption that the displacement is small, we shall return to this point later). The proportionality constant is the force constant of the spring and the expression for the opposing force is

$$F = -fx \qquad (1.5)$$

The displacement, x, of the masses from their equilibrium position, r_e, to some value r, is given by

$$x = r - r_e \qquad (1.6)$$

If the masses m_1 and m_2 represent atoms which are different they must attract the electrons constituting the bond to different extents and the molecule will have a permanent dipole moment. If the two masses represent identical atoms the diatomic molecule will have no permanent dipole moment. You may have met the concept of a dipole moment. If not, in brief, this is the product of the magnitude of the localized electrical charges within any molecule and the distance separating the positive and negative components of these charges. In other words dipole moment is the magnitude of electrical charge multiplied by the distance separating opposite polarities. Fig. 1.2a represents a classical description of the interaction of e.m. radiation with a diatomic molecule. The dipole moment is represented by an electric vector p.

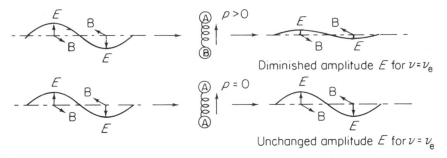

Fig. 1.2a. *Classical model of the interaction of e.m. radiation with a diatomic molecule*

A mechanism for displacing one of the masses relative to the other is by coupling the electric vector of the incident radiation to the electric vector of the molecular dipole moment as shown in Fig. 1.2a.

If the frequency, ν, of the incident radiation is equal, or to take a musical analogy is tuned, to a natural frequency of vibration of the molecule then the electric vector, E, of the incident radiation will couple with the dipole moment, p, and induce the molecule A-B to vibrate. Energy is taken up by the molecule from the incident radiation leaving a hole or absorption band in the spectrum. If the molecule has no permanent dipole moment (A-A) no electrical coupling can take place and no absorption takes place by this mechanism.

How may the natural frequency of vibration be calculated? We have considered the restoring force opposing displacement of the bond in terms of Hooke's law. This force is also given by Newton's second law of motion and the solution of this equation is a well known problem in calculus. If you have not met it before the answer is that the equilibrium frequency for a simple harmonic oscillator is given by

$$\nu_e = \frac{1}{2\pi} \sqrt{\frac{f}{\mu}} \tag{1.7}$$

where

$$\frac{1}{\mu} = \frac{1}{m_1} + \frac{1}{m_2} \tag{1.8}$$

Do you find the classical explanation of the vibrational spectrum of a diatomic molecule convincing? If so it is a fairly simple extension to a polyatomic molecule. Simple mechanics tells us that a system of masses joined by springs has a number of fundamental modes of vibration each of which has a particular natural frequency. Consider an oscillator such as the electric vector of e.m. radiation coupled to a system of masses such as a polyatomic molecule. By scanning through a range of frequencies some may be 'tuned' to the various fundamental modes of vibration by virtue of a change in dipole moment associated with that vibration. It is understandable that a series of absorptions take place for a polyatomic molecule rather than just one as in the case of a diatomic molecule because as we scan through a range of frequencies, radiation is absorbed each time we 'tune-in' or 'come into resonance' with the natural frequency of a fundamental mode which is capable of dipolar interaction.

SAQ 1.2b   Assuming the equilibrium frequency of the $H^{35}Cl$ molecule corresponds to 2886 cm^{-1} calculate the force constant of the $H^{35}Cl$ molecule in units N m^{-1} using the integer isotopic relative atomic mass value A_r (^1H) = 1, A_r (^{35}Cl) = 35.

SAQ 1.2c   Assuming the equilibrium frequency of $^{12}C^{16}O$ corresponds to 2143 cm^{-1} calculate the equilibrium frequency of $^{13}C^{16}O$ using integer mass numbers.

SAQ 1.2c

SAQ 1.2d   Which of the molecules listed below absorb infrared radiation?

H_2, HD, D_2, HF, F_2.

SAQ 1.2e   List the following molecules in decreasing order
of equilibrium vibration frequency

IBr, HI, IF, ICl.

SAQ 1.2f   List the following in increasing order of number of probable fundamental absorption bands
in the infrared

N_2 BF_3

SAQ 1.2f

If you accept a classical explanation for the interaction of e.m. radiation with the process of vibration for diatomic or polyatomic molecules what about the process of rotation? This is considered in the next section.

1.2.2. Classical Theory of Rotational Interactions

As in the case of vibrational interactions, electrical interactions are also feasible between e.m. radiation and dipolar changes in molecules arising from rotations. An important difference worth stressing from the outset is that free rotations are normally only possible in the gas state whereas vibrations can take place in all states of matter. The word 'normally' implies hidden caveats. In limited cases there is evidence for a degree of rotation in certain condensed states and we will consider some of this evidence at a later stage.

Take the same hypothetical diatomics as in the recent vibrational considerations. The observations now are that a heteronuclear molecule (A-B) absorbs not at one frequency as in the vibrational region but as a whole series of sharp lines in the rotational region. It is further observed that a homonuclear diatomic (A-A) does not absorb in the rotational region. A model to describe these interactions in terms of rotations is shown in Fig. 1.2b

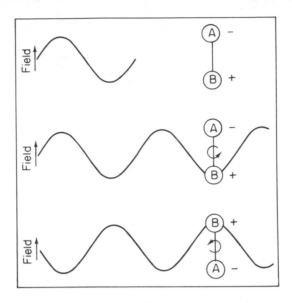

Fig. 1.2b.   *Classical model of a linear rigid rotor*

The model explains why a permanent dipole moment is essential before interactions can take place. A dipole moment can couple with an electric vector. If there is no dipole moment there is no coupling. This argument applies equally to the rotational spectrum of any polyatomic molecule. Hence spectra are only observed in the rotational region from gaseous molecules with permanent dipole moments. By contrast vibrational spectra may be observed in the infrared region from molecules in any phase which do not possess permanent dipole moments providing certain vibrations are associated with a change in dipole moment, which allow dipolar coupling between the molecule and the e.m. radiation.

Although the activity of rotational spectra is understandable, the sharp discrete structure is not explicable in classical terms. This discrete structure implies that the molecule can only rotate at certain discrete frequencies. We can explain the vibration of this molecule at a single vibrational frequency in classical terms but there is no obvious reason why it cannot rotate at a wide range of angular speeds corresponding to being induced to whiz round by e.m. radiation of a wide range of frequencies. The sharp line features are also apparent in other heteropolar diatomics and polar polyatomics. This line feature is one of a number of examples of the inadequacies of the classical theory.

SAQ 1.2g   Tick the viable combinations for spectral activity in the following table.

| | Permanent dipole | | No permanent dipole | |
	Vibrat.	Rotat.	Vibrat.	Rotat.
Solid				
Liquid				
Gas				

1.2.3. Classical Potential Energy Functions for Molecular Vibrations

So far we have only considered the frequency of absorption of vibrational and rotational processes. We also need to know how the potential energy changes with displacement for these processes. This is a very important relationship so it is worth deriving it in terms of simple calculus expressions for one particular case.

The vibrational potential energy, E_x, for a diatomic molecule is the force opposing the displacement multiplied by the displacement. This can be expressed as an indefinite integral

$$E_x = -\int F \, \mathrm{d}x \qquad (1.9)$$

If the diatomic molecule is assumed to be a harmonic oscillator
Hooke's Law may (Eq. 1.5) be applied which permits the integral
to be solved.

$$E_x = \int fx \, dx = \frac{1}{2} fx^2 + \text{Const.} \qquad (1.10)$$

If known values of f and r_e are substituted in the above equation
for E_x you can readily plot the vibrational potential energy of a
diatomic molecule as a function of x by selecting suitable values of
r. Before proceeding further it is worth doing this.

∏   For HCl, assume $f = 500$ N m^{-1} and $r_e = 0.130$ nm. Com-
        plete the following table and plot the results graphically

r	0.100	0.110	0.120	0.130	0.140	0.150	0.160
x	−0.03	−0.02	−0.01				
E_x	225	100	25				

The calculation is straightforward and proceeds as follows:

It is convenient to assume that when $x = 0$ then $E_x = 0$, and hence
the constant in Eq. 1.10 is also zero.

In all potential energy expressions the choice of the datum is arbitrary but it is essential to define the value for consistency.

For the data in the first column we get

$$E_x = 0.5 \times 500 \times (-0.03)^2$$

$$= 0.225 \text{ N m}^{-1} \text{ (nm)}^2$$

You may recall that N = J m^{-1}

$$\therefore \quad E_x = 0.225 \text{ J m}^{-1} \text{ m}^{-1} \text{ nm nm}$$

$$= 0.225 \times 10^{-18} \text{ J } 10^{-18}$$

$$E_x = 225 \times 10^{-21} \text{ J}$$

The table can be completed in a few seconds (I have done all the real calculating), and the graph in Fig. 1.2c completed.

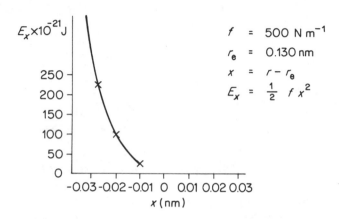

$$f = 500 \text{ N m}^{-1}$$
$$r_e = 0.130 \text{ nm}$$
$$x = r - r_e$$
$$E_x = \tfrac{1}{2} f x^2$$

Fig. 1.2c. *Classical potential energy function for HCl*

I hope you have ended up with a smooth symmetrical parabola. This is a result of assuming that Hooke's Law is obeyed at all values of x, and it is the classical potential energy function for a harmonic oscillator.

Look at Fig. 1.2d. and particularly at the case of the harmonic oscillator (i). I have plotted the full parabola and added some horizontal lines. The significance of these lines will be explained within the next few pages so just concentrate on the parabola for the moment.

The function suggests that if x is increased indefinitely E_x will increase indefinitely. Physical considerations tell us that when x reaches a sufficiently large value the bond will break and further increase in x leads to a value of E_x which tends to an upper limit. What happens to E_x when x is decreased to below zero? The function suggests E_x will increase in the same way for both negative and positive values of x by a parabolic curve. Physical considerations suggest that for negative values of x between $r = r_e$ and $r = 0$ there will be intense repulsion between the atomic nuclei

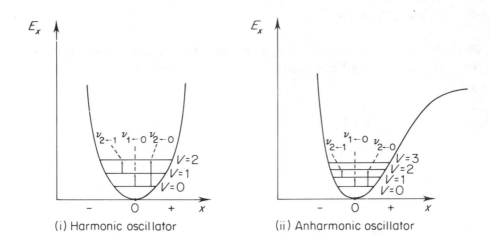

(i) Harmonic oscillator   (ii) Anharmonic oscillator

Fig. 1.2d.   *Potential energy functions of a diatomic molecule*

constituting the molecule and the value of E_x will increase more
sharply than the parabolic potential energy function suggests. A revised function based on these physical consideration is shown in
Fig. 1.2d(ii). You may note that for positive values of x, E_x increases less steeply than the parabolic or harmonic function, but
for negative values of x, E_x increases more steeply than the simple
function.

This new potential energy function differs from the one you and I
plotted in that it no longer corresponds to simple harmonic motion
and therefore may be termed *anharmonic*. One way of representing
an anharmonic function is by a power series

$$E_x = \frac{1}{2} f_1 x^2 - \frac{1}{6} f_2 x^3 + \frac{1}{24} f_3 x^4 - .. \qquad (1.11)$$

Here a series of anharmonic force constants exist f_1 f_2 f_3 etc. The
value of the numerical coefficients make sense if you take successive
higher differentials which correspond to different physical properties. It is clear that the negative value of f_2, the second term, en-
hances E_x for negative values of x but diminishes E_x for positive
values of x in accord with the physical considerations leading to
Fig. 1.2d(ii). Higher terms in this expression improve the correspondence with the realistic physical model.

Other mathematical expressions for vibrational potential energy have been proposed by Morse in 1929 and by Lippincot in 1955. These also reproduce Fig. 1.2d(*ii*) by using suitable parameters with particular physical significances. You may feel that it is a nonsense to apply Hooke's Law to molecules since the properties of chemical bonds are necessarily anharmonic and can in no way have their potential energy represented by a simple parabolic function. But before denigrating Hooke or his Law, look back and note the assumption made and consider whether within this assumption the Law applies to a diatomic molecule.

SAQ 1.2h

Sketch the potential function representing harmonic oscillation for a molecule with a low force constant and one with a high force constant. Suggest a typical pair of molecules in this category.

The underlying assumption in the expressions for the potential energy functions for vibrational energy (harmonic and anharmonic) is that molecular potential energy is a continuous variable. This assumption arises because we attribute to molecules, laws based on experience of macroscopic bodies such as large masses joined by springs, fly-wheels, engine governors controlling valves by centrifugal movements, etc. These models implicitly evoke mass itself as a continuous variable. At the molecular level it has long been accepted from the time of Dalton that mass was discrete and discontinuous. May it also be the case that at the molecular level energy itself is discrete and discontinuous? In which case our potential energy expressions are misleading in being expressed as continuous variables. The answer to this question emerges from observation of spectra.

1.3.   THE QUANTUM THEORY OF VIBRATIONAL AND ROTATIONAL SPECTROSCOPY

This Section aims to expose the limitations of classical explanations and the need for a general quantum mechanical theory leading to quantised energy levels.

In Section 1.2.1 we used a plausible model for understanding vibrational spectra of molecules in terms of simple classical mechanics of large vibrating objects but in Section 1.2.2 we noted that rotational spectra could, in no way, make sense when interpreted in terms of the mechanics of rotation of large bodies.

Possibly we were hasty or complacent in supposing we could understand the vibrational spectrum of, say carbon monoxide, as two masses joined by a spring. Let us have a closer look at the spectrum, first at low resolution in Fig. 1.3a and secondly at higher resolution in Fig. 1.3b. Do you see any anomalous features?

On the basis of a classical model the infrared spectrum of CO should consist of a single absorption band. At first sight, in Fig. 1.3a this is what we observe near 2143 cm^{-1}. On closer examination, in Fig. 1.3b when expanded at higher resolution, this band has structure

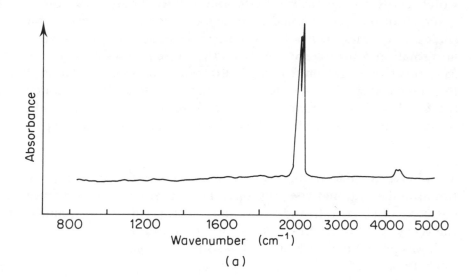

Fig. 1.3a.   *Infrared spectrum of CO gas at low resolution*

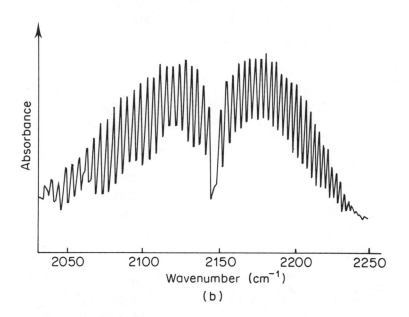

Fig. 1.3b.   *Infrared spectrum of CO gas at high resolution*

which turns out to have considerable detail. There is no obvious explanation for the detailed structure of this band on the ball and spring model and the answer lies in an interaction between vibration and rotation processes. Secondly there is a weak band at 4260 cm^{-1}. May this be an overtone? It is a feature of harmonic oscillators that they reveal second harmonics or overtones at twice the frequency of the first harmonic or fundamental. If so we would expect to find the overtone at 2 times 2143 = 4286. This is significantly higher than the observed value.

Thus classical ideas are inadequate in a number of respects for vibrational and rotational spectroscopy. The observation of vibrational spectra as in Fig. 1.3a and rotational spectra (not shown), and also the recognition of rotational structure to vibrational bands as in Fig. 1.3b was part of the experimental basis for the discovery of quantised vibrational and rotational energy levels.

SAQ 1.3a   Compare the infrared spectrum of CO as a gas (Fig. 1.3b) and as a solution in CCl4 (Fig. 1.3c). What does this suggest about the motion of the CO molecules in the two states?

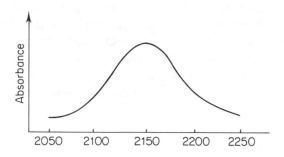

Fig. 1.3c. *Infrared spectrum of CO in solution in tetrachloromethane*

SAQ 1.3b

Fig. 1.3a and Fig. 1.3b shows the comparison of a portion of the spectrum of CO recorded on two different infrared spectrometers. What do you conclude about the difference in the operating condition of the two instruments?

1.3.1. Quantised Energy Levels

In Section 1.1.3 we noted the pioneering work of Balmer which established the existence of energy levels in the hydrogen atom. These are the simplest form of electronic energy levels in atoms or molecules. Thus the processes of vibration and rotation join the processes of electronic change and others listed in Fig. 1.1b as being represented in terms of quantised discrete energy levels E_0, E_1, E_2, etc. as in Fig. 1.3d.

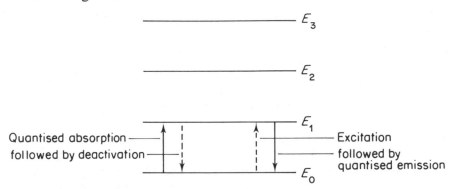

Fig. 1.3d. *Molecular energy levels*

Each atom or molecule in a system must exist in one or other of these levels. In a large assembly of molecules there will be a distribution of all atoms or molecules amongst these various energy levels.

These energy levels are a function of an integer (the quantum number) and a parameter associated with the particular atomic or molecular process associated with that state. Whenever an atom or molecule interacts with radiation a quantum of energy (or photon) is either emitted or absorbed. In each case the energy of the quantum of radiation must exactly fit the energy gap E_1-E_0 or E_2-E_1 etc. The energy of the quantum is related to the frequency (Section 1.1.3) by

$$\epsilon = h\nu \tag{1.4}$$

Hence the frequency of emission or absorption of radiation for a transition between the energy states E_0 and E_1 is given by

$$\nu = \frac{E_1 - E_0}{h} \qquad (1.12)$$

We will use the accepted convention for transitions from E_0 to E_1 as $E_1 \leftarrow E_0$ rather than $E_0 \rightarrow E_1$.

The concept of quantised energy levels thus explains the observation noted in Section 1.1 that the frequencies of bands in absorption spectra were the same as the frequencies of bands in emission spectra in the absence of decomposition or other energy transfer processes. Associated with the uptake of energy of quantised absorption is some deactivation mechanism whereby the atom or molecule returns to its original state; associated with the loss of energy by emission of a quantum of energy or photon is some prior excitation mechanism. Both these associated mechanisms are represented by dotted lines in Fig. 1.3d.

SAQ 1.3c

A typical energy change associated with a rotation transition is 15×10^{-23} J.

For a vibration transition it will be in the region of 100 times greater.

Calculate the energy in joules, of the vibration transition 2143 cm^{-1} for a CO molecule.

SAQ 1.3c

Let us now examine an energy level diagram for the carbon monoxide molecule which helps to explain the features of the infrared spectrum of CO shown in Fig. 1.3a and Fig. 1.3b. We will then justify the validity of the diagram and the associated ideas using wave mechanics.

Infrared spectroscopy is concerned with vibrational and rotational energy changes and such an energy diagram in Fig. 1.3e shows a number of quantised energy states for both vibrations and rotations.

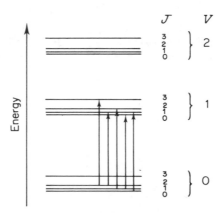

Fig. 1.3e.   *Vibration and rotation energy levels for a diatomic molecule such as CO*

The diagram shows a stack of vibrational levels. Only three are shown for simplicity and they are characterised by V which takes values 0, 1, 2, ... You will notice that each of these vibrational energy states has associated with it a stack of rotational energy states. Only four are shown in order to keep the diagram clear. These rotational energy levels are characterised by J which takes values 0, 1, 2, 3, ...

We can now explain the CO spectrum, as follows:

The main absorption band at 2143 cm^{-1} is due to a vibrational energy transition from energy level $V = 0$ to level $V = 1$, $\nu_{1 \leftarrow 0}$.

The weak absorption band at 4260 cm^{-1} is due to a vibrational energy transition, $\nu_{2 \leftarrow 0}$.

The detailed structure associated with these spectral bands is due to the rotational energy changes that take place. Hence when a transition takes place from the $V = 0$ level, different molecules will have slightly different rotational energies, some occupying level $J = 0$, others $J = 1$ and so on. Fig. 1.3e shows five examples of $1 \leftarrow 0$ vibrational transitions.

Bearing in mind that this energy level diagram only shows a few levels in each stack you will appreciate that very many more transitions are possible. Over 50 peaks due to over 50 different energy transitions make up the 2143 cm^{-1} infrared absorption band of CO!

Can we explain these spectral observations rigorously? Yes we can using wave mechanics. The theory however, is somewhat abstract and complex, and we will only take a quick look before moving on.

A general description and basis for calculating the properties of all energy states in atoms and molecules is by the methods of wave mechanics developed by Planck, de Broglie, Einstein and Schrödinger. Those interested in pursuing the mathematical details are advised to consult standard textbooks. For immediate purposes, it will suffice to outline the formulation of the general problem in terms of the mathematical model, the input to the model, and the solution which provides the output. The particular results which come out for rotation and vibration states will be listed.

The mathematical model consists of the appropriate potential energy expression which is substituted as an input to the Schrodinger wave equation. A wave function, ψ, is then found which satisfies the equation. Only certain allowed functions (eigenfunctions) satisfy the equations leading to allowed values (eigenvalues) corresponding to particular quantised energy states.

Let us take a short cut and consider the solution of the Schrodinger wave equation for four different systems. Look at Fig. 1.3f. You will see that we consider a diatomic molecule, A-B, rotating rigidly and non-rigidly, and the same molecule vibrating harmonically and anharmonically. Now examine the harmonic oscillator system in more detail. You will notice that we have used the classical description of harmonic oscillation to give us an expression for potential energy which you will recognise as Eq. 1.10.

The ensuing wave mechanics gives us the equation for calculating the energy of the quantised energy levels.

$$E_v = h\nu_e(V + \tfrac{1}{2}) \tag{1.13}$$

The energy is a function of an integer V which is the vibrational quantum number and ν_e which is the frequency of the molecular vibration and hence the frequency of the e.m. radiation with which the molecule can interact.

When $V = 0; E_0 = \tfrac{1}{2} h\nu_e$

   $V = 1; E_1 = \tfrac{3}{2} h\nu_e$

   $V = 2; E_2 = \tfrac{5}{2} h\nu_e$

Note that the energy levels are 'equally spaced'.

I would now like you to go back and look at Fig. 1.2d and examine the case of the harmonic oscillator. The three energy levels

System	Model	Potential energy function	Quantised energy levels	Wave funtion
Rigid rotor	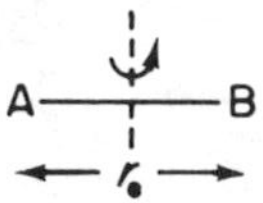 A–B	$V = 0$	$E_J = hBJ(J+1)$	ψ (rigid)
Non-rigid rotor	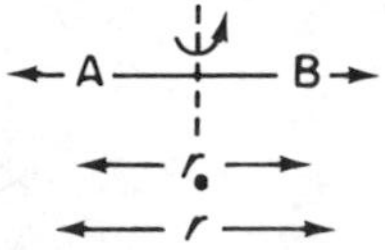 A–B	$V = E_x = \frac{1}{2}fx^2$	$E_J = hBJ(J+1) - hDJ^2(J+1)^2 + \dots$	ψ (non-rigid)
Harmonic oscillator	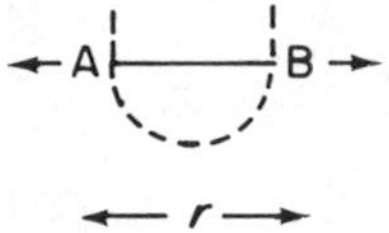 A–B	$E_x = \frac{1}{2}fx^2$	$E_v = h\nu_e(V+\frac{1}{2})$	ψ (harmonic)
Anharmonic oscillator	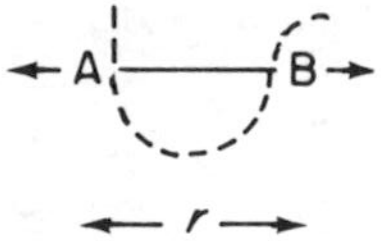 A–B	$E_x = \frac{1}{2}f_1x^2 - \frac{1}{6}f_2x^3 + \dots$	$E_v = h\nu_e(V+\frac{1}{2}) - h\nu_e x_e(V+\frac{1}{2})^2 + \dots$	ψ (anharmonic)

I = moment of inertia about perp. axis passing through the centre of gravity

f = force constant of the A—B bond

$x = r - r_e$

f_1, f_2 = coefficients in the anharmonic potential energy function

$B = h/8\pi^2 I$ = rotational constant

J = rotational quantum number

D = centrifugal distortion constant

ν_e = equilibrium frequency

v = vibrational quantum number

x_e = anharmonicity constant

Fig. 1.3f. *Parameters associated with rotating and vibrating diatomic molecules*

have been drawn on the classical potential energy diagram that we considered previously.

∏   If the $\nu_{1 \leftarrow 0}$ transition for CO is 2143 cm^{-1} what value does this model predict for the $\nu_{2 \leftarrow 0}$ transition?

4286 cm^{-1}, because the levels are equally spaced. The actual value is 4260 cm^{-1} and harmonic oscillation does not accurately describe the vibrational behaviour of CO.

If you now look at the case of the anharmonic oscillator as depicted in Fig. 1.3f and Fig. 1.2d you will see that the energy levels are not equally spaced. The extent of the inequality, that is the extent of the anharmonic behaviour, is given by the anharmonicity constant. We can now explain why $\nu_{2 \leftarrow 0} < 2 \times \nu_{1 \leftarrow 0}$.

We could go further and calculate intensities of absorption bands and predict and explain spectral features exhibited by many different molecules. We are not going to get involved with too much detail, but there are more fundamental aspects of infrared spectroscopy that need to be considered before leaving this Part of the Unit.

1.4.   DISTRIBUTION AND LIFE-TIMES OF ENERGY STATES

This section aims to explain the significance of the Maxwell–Boltzmann distribution of molecules amongst energy levels and to provide some measure of the life-time of excited states of energy levels in relation to emission and absorption in the infrared and microwave regions.

You will have gathered from the previous Section that any molecule has a number of stacks of energy levels, each stack corresponding to a particular process such as electronic, vibrational or rotational change. (Here reference to electronic energy levels is included for completeness but only vibrational and rotational levels are dealt with in any detail). For large molecules there are a number of possibilities for each type of process. Thus HCl has a number of stacks of electronic energy levels but only one stack of vibrational levels and one stack of rotational levels. H_2O has, in addition to its electronic levels, three stacks of vibrational levels and three stacks of rotational levels. This is because there are three fundamental modes of vibration and three modes of rotation.

Fig. 1.3e shows one particular stack of energy levels. For each stack of this type each molecule must exist in one or other of these energy levels. Can you make the mental jump from an isolated molecule to a large assembly or system of molecules? There will be a distribution of all the molecules between these various energy levels. The relative population of molecules, (N_i/N_j) in any two levels E_i and E_j is given by the Maxwell–Boltzmann equation.

$$\frac{N_i}{N_j} = \frac{g_i}{g_j} e^{\frac{-\Delta E}{kT}} = \frac{g_i}{g_j} e^{\frac{-h\nu}{kT}} \tag{1.14}$$

where g_i and g_j are the number of permitted states with energy E_i and E_j (the degeneracy of the state), ΔE is the difference in energy between the states $(E_i - E_j)$, k is the Boltzmann constant and T is the absolute temperature. It follows that if T is very small or ΔE is very large the higher energy states will be almost completely depopulated. Conversely if T is large or ΔE is small the higher energy states will be highly populated. this may not be too apparent from the expression so we will carry out some calculations to illustrate this principle. Calculations involving exponential terms are easier with suitable pocket calculators. It is worth getting familiar with these functions because they have important physical significance in a number of areas, including understanding spectra!

SAQ 1.4a | The table below lists data for the relative population of two energy states E_0 and E_1 for various temperatures and for energy separations represented by spectral transitions in wavenumbers, cm^{-1}. $\longrightarrow$

**SAQ 1.4a
(cont.)**

T/K $\bar{\nu}$/cm^{-1}	10	30	100	300	1000	3000
10	0.24	6×10^{-7}	0	0		0
100	0.87	0.65	0.24		6×10^{-7}	0
300	0.96	0.87	0.62	0.24	8×10^{-3}	6×10^{-7}
1000	0.99	0.96	0.87	0.65	0.24	

(*i*)   Three gaps have been left for you to complete; calculate the relative population of the higher to the lower level assuming the degeneracy of each level is the same. You might like to use the same equation that I used in calculating the listed values:

$$\frac{N_1}{N_0} = \exp\left(\frac{-(E_1 - E_0)}{kT}\right) =$$

$$\exp\left(\frac{-hc\bar{\nu}_{1 \leftarrow 0}}{kT}\right) = \exp\left[-1.438\left(\frac{\nu_{1 \leftarrow 0}}{T}\right)\right]$$

(*ii*) Does the population of the excited state E_1, increase or decrease as the temperature is raised?

(*iii*) Is the population of the excited state greater or smaller for the 1000 cm^{-1} transitions compared with those at 30 cm^{-1}, at a given temperature?

(*iv*) How do you think the intensity of a transition of the type $E_2 \leftarrow E_1$ will differ for the two cases:

$$T = 100 \text{ K}, E_1 \leftarrow E_0 = 10 \text{ cm}^{-1}$$

$$T = 100 \text{ K}, E_1 \leftarrow E_0 = 300 \text{ cm}^{-1}?$$

SAQ 1.4a

1.4.1. Life-times of Excited States

It is tempting to assume that when a sample is placed in the beam of a spectrometer this is a static system as compared with a more obviously dynamic system such as when we mix two reactive chemicals. Perhaps we should consider the insertion of a sample in a spectrometer beam as a particular kind of reaction between a photon and a molecule as shown in Fig. 1.4a.

Here the transition from the lower to the upper level is shown together with the resulting absorption band.

If the molecule remained in the upper energy state indefinitely the absorption band would supposedly become weaker with time as photons saturated the sample with energy. Since time-dependent changes are observed only infrequently it follows that excited energy states have a short natural life-time.

If the source of radiation is sufficiently powerful the higher energy

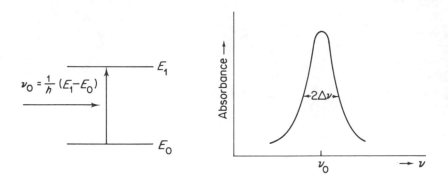

Fig. 1.4a.   *Typical absorption transition and its observed absorption spectrum*

state may become saturated. This leads to power saturation broadening of the band. This phenomena may occur in the microwave or radiowave regions unless the power of the source is kept below a critical level. In regions to higher frequency, power saturation broadening is only observed when intense sources such as lasers are used.

There is a relationship between the life-time of any excited state and the band-width of the absorption band associated with the transition to this excited state. This relationship is a consequence of the Heisenberg uncertainty principle which states that at the microscopic level the product of the uncertainty of any two conjugate variables such as time and energy is constant. The particular form in the present context is that the product of the life-time of the excited state (τ) and the uncertainty of the energy of the state (δE) is greater than or equal to $h/2\pi$

$$\tau \times \delta E \geq h/2\pi \qquad (1.15)$$

since

$$\delta E = h(2\Delta\nu) \qquad (1.16)$$

$$\tau \times \Delta\nu = \frac{1}{4}\pi \tag{1.17}$$

where $\Delta\nu$ is the half-width of the band at half-height as shown in Fig. 1.4a.

It follows that if the life-time is very large $\Delta\nu$ is very small and the band is very sharp. Conversely if the life-time is very short $\Delta\nu$ is relatively large and the band is broad. If spontaneous decay of an excited state according to its natural life-time is assumed to be the main factor controlling band-width, then the band-width in different regions of the spectrum can indicate the life-time of the excited states in these regions. There are, however, several other factors contributing to spectral band-width.

In the gas phase bands are intrinsically sharp and some broadening occurs mainly from Doppler effects and pressure effects. There is, however, no need for us to consider these effects further.

In later Parts of the Unit virtually all the spectra you meet will be those of compounds in either the liquid or solid phase. Rotational fine structure, such as that met earlier in the gas phase spectrum of CO, will not be resolved because of the appreciable band broadening by inter-molecular interactions.

1.5. VIBRATIONS OF POLYATOMIC MOLECULES

Previous Sections applied simple classical and quantum ideas to explain spectra, with emphasis on infrared spectra of diatomic molecules. The present Section aims to extend the explanations to polyatomic molecules.

In Section 1.2.1 we noted that in the classical description of diatomic molecules the wavenumber of the absorption band in the infrared spectrum was related to two masses joined by a spring. By simple mechanical analogy this system gives rise to a single fundamental mode of vibration which is associated with the prominent band in the spectrum.

The problem obviously becomes more difficult for a polyatomic molecule since there is more than one spring-like bond and at least three atoms. For example we could consider the infrared spectrum of compounds from water with three atoms to a protein like lysozyme with many thousands of atoms. What are the modes of vibration in such molecules?

1.5.1.   Normal Modes of Vibration

A polyatomic molecule can be looked upon as a system of masses joined by bonds with spring-like properties. The position of each of these atoms can be described using three numbers, indicating position relative to a set of axes at right angles (the X, Y and Z axes).

An atom has three degrees of freedom. This means it can move relative to each of these axes. This is known as translational movement. Molecules have many more degrees of freedom since their many atoms can move relative to each other.

∏   How many degrees of freedom will a diatomic molecule
       have?

It is reasonable to assume that it will have twice as many as one atom, that is it will have six rather than three.

∏   What form will these six degrees of freedom take?

A diatomic molecule such as H—Cl, like any other body, can only have three degrees of translational freedom. What are the other three? The answer is rotational and vibrational degrees of freedom. For a diatomic molecule, or any other linear molecule, rotation is only meaningful about two axes since rotation along the third axis which is the axis of the molecule does not have any meaning since it involves no energy change. Two degrees of freedom are therefore used for rotation. The remaining degree of freedom is the vibration of the molecule by virtue of the opposing motion of the two atoms associated with stretching and compressing the bond.

Degrees of Freedom	CO_2 linear	H_2O non-linear
Translational	3	3
Rotational	2	3
Vibrational	$3N - 5$	$3N - 6$
Total	$3N$	$3N$

Fig. 1.5a. *Degrees of freedom for polyatomic molecules*

For polyatomic molecules containing many (N) atoms we can distinguish two groups: linear and non-linear. CO_2 and H_2O are simple examples. Both linear and non-linear molecules will have three degrees of translational freedom. Linear molecules will have only two degrees of rotational freedom (as above for the diatomic molecule) while non-linear molecules will have three since it is now meaningful to rotate about all three axes. *Hence linear molecules will have $3N - 5$ and non-linear molecules will have $3N - 6$ degrees of vibrational freedom* (See Fig. 1.5a).

SAQ 1.5a

How many vibrational degrees of freedom do the following molecules possess?

(*i*) Ammonia (NH_3) Answer ...

(*ii*) Ethene ($CH_2{=}CH_2$) Answer ...

(*iii*) Propyne ($CH_3C{\equiv}CH$) Answer ...

SAQ 1.5a

Our system of masses and springs can only vibrate in certain modes known as 'normal' or 'fundamental' modes of vibration. Each mode corresponds to a degree of vibrational freedom and has a characteristic frequency and form of vibration. By coupling a mechanical system to an oscillator of variable frequency it is possible to observe the successive excitations of each mode within the mechanical system through the full range of frequencies. It is a fundamental mechanical phenomena that an oscillator will only induce sympathetic vibrations in any system at certain natural frequencies. It is sometimes said that at these particular frequencies the oscillator and the system are in 'resonance'.

SAQ 1.5b   Can you think of examples of 'resonance' in the areas of civil engineering, electrical engineering or acoustics?

SAQ 1.5b

Thus returning to the microscopic world of atoms and molecules; each of the $3N - 6$ (or $3N - 5$) normal or fundamental modes of vibration involves internal motions of atoms in which all the atoms move in phase and with the same frequency but with different amplitudes and in different directions. The amplitude and direction of each atom may be represented by a displacement vector. The various displacements of the atoms in a given normal mode of vibration may be represented by a linear combination of the displacements of all atoms which is known as the normal coordinate of that mode. These may be represented pictorially for selected molecules. Let us do this for two simple cases.

$$\longleftarrow \text{H--Cl} \rightarrow \qquad \leftarrow \text{C--O} \rightarrow$$

In any vibration the centre of gravity remains constant and the displacement of the lighter atoms are greater than for the heavier atoms as shown for HCl and CO.

Whereas a diatomic molecule has only one mode of vibration which corresponds to a stretching motion, a non-linear BAB type triatomic molecule has three modes two of which correspond to stretching motions and the remainder corresponds to a bending motion. A linear type triatomic has four modes two of which have the same frequency and are said to be *degenerate*. These are shown in Fig. 1.5b.

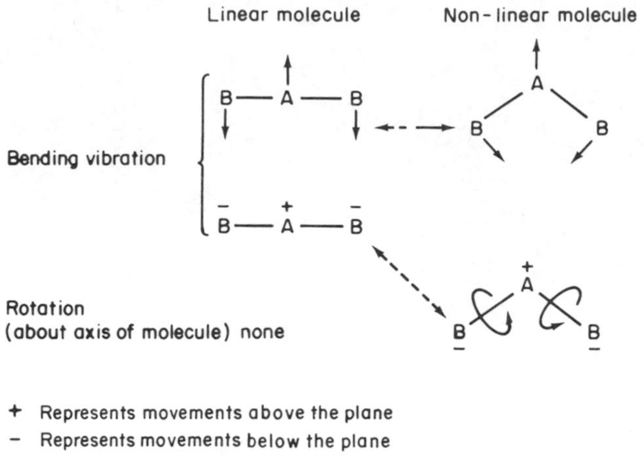

Fig. 1.5b. *Relationship between degenerate bending vibrations and rotational motions in triatomic molecule, BAB*

Π   How do we represent vibrational motions?

As arrows in the plane (if we knew the amplitude of the vibration we would represent it by the length of the arrow). As + or − above and below the plane.

To illustrate the concept of degeneracy take a model of a flexible B-A-B system and consider the rotational motion about the B-A-B axis of the molecule as the B-A-B angle changes from 180 to 120. The energy of rotation is zero initially but increases as the angle decreases.

This accounts to a gain of one degree of rotational freedom and a loss of one degree of vibrational freedom. Can you see in Fig. 1.5b that a particular rotational motion in a non-linear molecule is equivalent to a particular bending vibration in a linear molecule? Also that this bending vibration corresponds to another bending vibration at right angles. These two vibrations are called a *degenerate* pair and will give rise to only one spectral band which will be described as *doubly degenerate*.

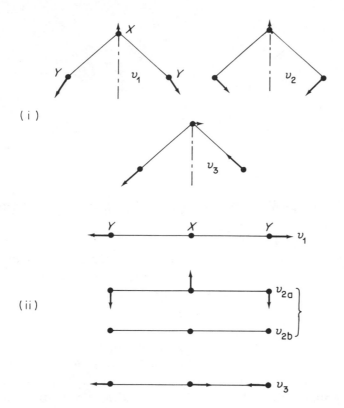

Fig. 1.5c. *Normal vibrations of a non-linear and a linear triatomic molecule, YXY*

If you can accept this difference between a linear and a non-linear molecule compare their vibrations in Fig. 1.5c. In each case there are two modes corresponding to a symmetric and an antisymmetric stretching mode and a third mode corresponding to a bending mode.

The displacement vectors depend on the particular molecules and Fig. 1.5d shows the differences in these vibrations for H_2O and SO_2.

Although the amplitudes differ considerably each molecule has a well defined symmetric stretching mode, an antisymmetric stretching mode and a bending mode. The pairs of O—H and S—O bonds are said to be coupled into symmetric motions at lower wavenumber, and into antisymmetric motions at higher wavenumber.

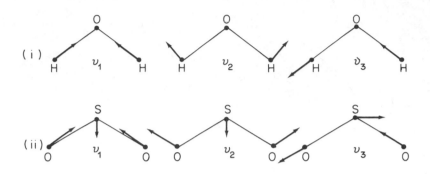

Fig. 1.5d. *Actual form of the normal vibrations of H_2O and SO_2*

∏ Where would you expect the O—H-stretching wavenumber values in an alcohol, given that the symmetric and antisymmetric stretching modes occur at 3652 and 3756 cm^{-1} respectively for water in the gas phase?

It is reasonable to assume this is the mean of the symmetric and antisymmetric modes $(3756 + 3652)/2 = 3704$ cm^{-1}. You will observe that the OH stretching mode occurs to slightly lower values than this typically near 3600 cm^{-1}. The reduction is attributable to the electron donor properties of the alkyl group. This mode is also reduced in the condensed state and also as you will see in Part 5 strongly influenced by hydrogen bonding.

CO_2, H_2O and SO_2 are examples of molecules which are symmetrical about the central atom. Their vibrations are influenced by this symmetry which also determines the activity or inactivity of vibrations in the infrared and Raman spectra.

∏ In Fig. 1.5c which vibration would you expect to be inactive or absent from the infrared spectrum?

The answer is ν_1 for the linear structure since this is the only vibration in the Figure for which the resultant dipole moment does not change during the vibration (see Section 1.2.1). The subject of activities and intensities of vibrations will be considered in Section 1.5.2.

If the molecule has different terminal atoms such as HCN, ClCN or ONCl then the two stretching modes are no longer symmetric and antisymmetric vibrations of similar bonds but will have varying proportions of the stretching motion of each group. In other words the amount of *coupling* will vary. This effect is shown in Fig. 1.5e for HCN and ClCN. You may notice that the H—C and C≡N stretching motions are not significantly coupled to the other motions and may give rise to characteristic vibrations for these groups. For simplicity only one component of the bending vibration is described.

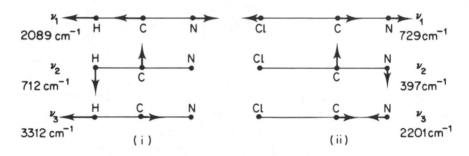

Fig. 1.5e. *Actual form of the normal vibrations of (i) HCN and (ii) ClCN*

SAQ 1.5c   List the characteristic group frequencies in the following with reference to Fig. 1.5e.

HCN ...

ClCN ...

To what do you attribute any characteristic features of these values?

SAQ 1.5c

If we consider the possible molecular shapes for tetra-atomic molecules we find these are considerable. One common kind is of the type AB_3 which may be planar as in the case of BF_3 or pyramidal as in the case of NH_3. Other types are AB_2C, or ABCD possessing a miscellany of shapes which determine, and may be deduced from, spectral and other data. It should be apparent that the vibrational spectrum of a polyatomic molecule has many features. As the number of atoms becomes large the number of fundamental vibrations may increase nearly three-fold. As in the case of diatomic molecules, weak overtones may be observed with wavenumber values which are approximately twice those of the corresponding fundamental. Also sum (or difference) combinations may appear at wavenumber values which are approximately the sum (or difference) of wavenumber values of the component fundamentals. These are all examples of binary combinations. Very weak ternary combinations may also be formed in similar ways with an increase in the number of possible combinations.

Hence an infrared spectrum can be highly complex with many fundamental and combination bands. Some of these will be weak, others will overlap each other and some may be below the wavenumber range of the spectrometer used. In view of this complexity how may we proceed? The answer is that only rarely is a full interpretation carried out; more often only a partial interpretation of the spectrum is necessary for two reasons.

Firstly certain chemical groups have masses and force constants which are sufficiently different from those in adjacent groups to create characteristic group frequencies or vibrations at wavenumber values which can be compared with data in correlation tables or reference spectra. Specific groups can therefore be identified from wavenumber values which are typically between 1500 and 4000 cm^{-1}.

A detailed treatment of infrared group frequencies will be given in Part 4 of this Unit.

Secondly the complexity of a spectrum is likely to be a unique pattern which may identify an unknown material by comparison with your own reference spectra or those in commercial spectral collections. The complexity arises from coupling of vibrations over a large part of, or over the complete molecule. Such vibrations are called *skeletal* vibrations. Bands associated with skeletal vibrations are likely to conform to a pattern or 'fingerprint' of the molecule as a whole rather than a specific group within the molecule and tend to occur between 50 and 1500 cm^{-1}.

Whether you wish to carry out a detailed interpretation in relation to some aspect of structure determination, a partial interpretation in relation to identification, or some quantitative analysis from band intensity measurement the factors governing intensity of infrared bands are as important as the factors governing frequencies. Infrared intensities will be considered in the same way as infrared frequencies in terms of simple molecules, as a basis for an adequate understanding of many applications of vibrational spectroscopy which are treated in the remaining Parts of this Unit.

1.5.2. Intensity of Infrared Bands

In Section 1.5.1 we considered molecular vibrations without much regard to molecular spectra. The reason for this apparent departure from our earlier premise that molecular spectroscopy is an experimentally based subject is that it is easier to swallow the concepts of vibrational frequencies and vibrational intensities if treated separately.

In Section 1.2.1 we devised a very simple classical model for explaining the frequency and intensity of a diatomic molecule. Have another look at Fig. 1.2a in which the diatomic molecule has a natural frequency of vibration which as shown by Eq. 1.7 is determined by the force constant of the bond and the masses of the atoms. You will notice by looking around you that vibrations do not occur spontaneously, the system has to receive energy from an external source.

Try crossing the Severn Bridge or the Golden Gate Bridge in a force ten gale!

Unless you have a deep love of the forces of nature and total confidence in Civil Engineers you will wish you were back in the laboratory carrying out a less violent experiment in energy transfer! Is the analogy fair? The force of the wind may be constant or it may oscillate and it corresponds to mechanical movement of air. A bridge is a composite of atoms and polyatomic species with many modes of vibration. There are some spectacular films of consequences of vibrations of bridges induced by wind power.

To return to the molecular level in Fig. 1.2a we illustrated energy transfer by coupling between the electric vector of the e.m. radiation and an electric dipole in the molecule. If there is no electric dipole as in a molecule of type, A-A, no energy is absorbed. If a dipole exists as in A-B type molecules energy is absorbed when the frequency of the e.m. radiation and the frequency of the vibration coincide. We may say that the e.m. radiation now has something to catch hold of electrically to set the vibration in motion.

How do we extend this model to a polyatomic molecule of N atoms with up to $3N - 6$ fundamental or normal modes of vibration and possible weaker binary, ternary and higher combinations? We have described in Section 1.5.1 a number of vibrations which may in principle be induced by e.m. radiation when scanned through the range of frequencies possessed by the molecule. the condition for energy transfer is that there must be a *change of dipole moment during the vibration for the vibration to be infrared active*; by infrared active we mean for that vibration to interact with e.m. radiation of the same frequency leading to an absorption band in the spectrum. The larger the change in dipole moment the greater the intensity of the absorption band associated with the vibration.

∏ Which vibration(s) shown in Fig. 1.5c would you expect to be inactive and active respectively in the infrared spectrum?

The totally symmetric stretching mode of the linear molecule is inactive in the infrared, since the dipole moment is always zero and there is no change in dipole moment during the vibration. Can you see that all other vibrations shown in Fig. 1.5c(i and ii) give rise to changes in dipole moment? This change leads to absorption bands in the infrared spectrum at the frequency of these vibrations. Hence ν_1 for a linear centro-symmetric triatomic molecule is infrared inactive but ν_2 and ν_3 are infrared active. For a non-linear centro-symmetric molecule all three fundamental modes are infrared active.

For simple cases we may estimate likely activities of molecular vibrations and hence estimate whether particular vibrations are likely to lead to strong or weak absorption bands.

∏ Estimate the likely intensity of the following absorption bands

(i) C=C stretch 1,1-dichloroethene

(ii) C=O stretch in acetone

(iii) C=C stretch in trans 1,2-dichloroethene.

In (iii) there is no change in dipole moment and hence this vibration is inactive in the infrared. In (ii) there is a large dipole moment associated with the carbonyl group and a large change in this value during the vibration which is therefore associated with a characteristically strong carbonyl band. In (i) we could expect a dipole change somewhere between (ii) and (iii) leading to a medium intensity absorption band.

1.5.3. The Quantum Theory of Vibrational Spectra of Polyatomic Molecules

All we have said so far in Sections 1.5.1 about vibrations of poly-atomic molecules and in Section 1.5.2 about intensity considerations has rested on classical theory. It could have been presented 100 years ago or more, before the pioneers of modern quantum theory had revolutionised science. Vibrational spectroscopy is unique in being largely explicable in classical terms. You will meet many problems in which little or no quantum theory is necessary. When do we meet problems which really need quantum theory?

We met two such problems when we looked at the spectrum of CO in Section 1.3. The first concerned the rotational structure of the infrared absorption band, and the second concerned the nature of the overtone in terms of frequency and intensity. Furthermore in Section 1.4 we considered the distribution and life-times of energy states including those associated with vibrational energy.

For a diatomic molecule we considered a single stack of energy levels or states E_0, E_1, E_2, E_3, ... E_V etc each characterised by the vibrational quantum number 0, 1, 2, 3 ... V, etc. The magnitude of the *difference in energy* between two states is determined by measurement of the frequency of the band corresponding to a transition between these two states.

The energy of each state is predicted by solving the Schrödinger Wave Equation for a particular system. the solutions are listed in Fig. 1.3f and enable the frequencies to be predicted for a molecule treated as a simple harmonic oscillator or as an anharmonic oscillator.

For a diatomic molecule there is only one stack of vibrational energy levels as shown in Fig. 1.3d. For a polyatomic molecule there are $3N - 6$ stacks of vibrational energy levels for non-linear molecules and $3N - 5$ stacks for linear molecules. We may illustrate these stacks for the linear CO_2 molecule, Fig. 1.5f compares the classical and quantum descriptions and shows that two of the four fundamental vibrations are degenerate. This leaves three fundamental vibrational frequencies.

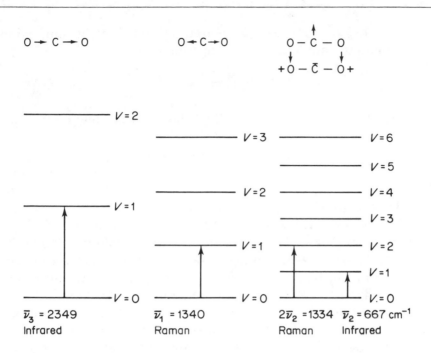

Fig. 1.5f. *Vibrational energy levels for CO_2. ν_1 symmetric stretching vibration, ν_2 asymmetric stretching vibration and ν_3 bending vibration*

∏   Why are only two of these observed in the infrared spectrum?

The symmetric stretching mode, ν_1, is inactive because there is no change in dipole moment associated with the vibration. You will note, however, that ν_1 is Raman active.

Raman spectroscopy is a technique which is complementary to that of infrared spectroscopy. We are not going to deal with it is this Unit but you should note its existence, and that it is an alternative technique for studying molecular vibrations and can sometimes provide information not available in an infrared spectrum. For example the symmetric stretching mode for CO_2 is Raman active.

Selection Rules. Quantum theory and wave mechanics enables the selection rules of transitions to be determined. For the simple harmonic oscillator model the selection rule is

$$\Delta V = \pm 1$$

For the anharmonic oscillator model the rules are less restrictive.

$$\Delta V = \pm 1 \pm 2 \pm 3 \text{ etc.}$$

What does this mean? In Fig. 1.5f we have shown the transition from $V = 0$ to $V = 1$ for ν_2 and ν_3. These correspond to $\Delta V = 1$. For ν_2 we have shown not only the transition for $\Delta V = 1$ but also for $\Delta V = 2$. The first overtone of the bending mode, $2\nu_2$ corresponds to $\Delta V = 2$ and turns out to be inactive in the infrared but active in the Raman leading to a strong feature in the Raman spectrum. By contrast, ν_2 the fundamental of the bending mode which corresponds to $\Delta V = 1$ turns out to be active in the infrared but inactive in the Raman. This illustrates the distinction between activity and selection rules. Further aspects of these results will be amplified later in this Unit.

Can you see it is important to retain the quantum, as well as the classical model of the vibrations of a polyatomic molecule? Without this we cannot understand anything about rotational structure and cannot have a clear appreciation of selection rules for fundamentals and overtones. Without the quantum model we cannot understand the energy of a vibrational state, the relative population of these states in terms of the Maxwell–Boltzmann law or the life-times of these states.

Notwithstanding those effects which are inexplicable without a quantum approach to vibrational spectroscopy, much of the infrared spectra of organic molecules is interpreted largely in terms of classical ideas of characteristic group frequencies and skeletal frequencies. Further aspects of this approach will be considered in Part 4 of this Unit.

SAQ 1.5d

Suggest the likely forms of the modes of vibrations of the formate ion. It will help to consider this system as an extension of a non-linear triatomic molecule which is illustrated in Fig. 1.5c.

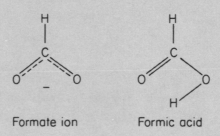

Formate ion   Formic acid

SAQ 1.5e   Predict the main features of the infrared spectrum of formic acid.

SAQ 1.5f   List five diatomic molecules which are infrared active and five which are infrared inactive.

SAQ 1.5f

SAQ 1.5g   State the rule governing the activity of any fundamental vibrational mode in the infrared region.

SAQ 1.5h   Indicate the factors influencing the intensities of
                absorption bands of C=O and C=C stretching
                modes.

Summary

This first Part of the Unit introduces ideas and theories fundamental
to the understanding of infrared spectroscopy.

Initially the electromagnetic spectrum is considered in relation to
various atomic and molecular processes. Interactions of infrared and
microwave radiations with matter are then dealt with by reference
to classical models which describe changes in dipole moment associated with molecular vibrations.

The limitations of classical models are exposed and quantum theory with an associated wave mechanical treatment is introduced. The treatment is illustrated with reference to simple diatomic and triatomic molecules and their infrared spectra.

All the main features of such spectra are accounted for before finally turning to polyatomic molecules.

The complexity of infrared spectra of polyatomic molecules is introduced along with the concept of characteristic group frequencies which play a major role in later Parts of the Unit.

Objectives

You should now be able to:

● understand the means by which electromagnetic radiation interacts with a vibrating molecule;

● predict the number of fundamental modes of vibration of a molecule;

● appreciate the factors governing the position and intensity of bands in an infrared spectrum;

● explain qualitatively all the features of an infrared spectrum of a diatomic molecule such as CO, by reference to classical and quantum mechanical models;

● appreciate the different possible modes of vibration of the type: stretching (symmetric and antisymmetric), bending, degenerate, characteristic group, coupled and skeletal.

2. Instrumentation for Infrared Spectroscopy

By now you should be familiar with the relationship between the *processes* of vibration in molecules and the *method* of infrared spectroscopy. We now consider the *instrumentation* associated with this method. I hope you have already recorded an infrared spectrum of at least one sample, because even such limited hands on experience will help you to appreciate the significance of much of what follows in this Part of the Unit.

Do you recall the General Introduction to the Electromagnetic Spectrum in Section 1.1 of Part 1? We demonstrated a visible spectrum using a glass prism as a dispersing element. Could we use a glass prism to obtain a spectrum beyond the red in the infrared? No, because glass is opaque to infrared radiation beyond 2.5 μm. But there are many optical materials, transparent to infrared radiation which have been used as prisms. Prism instruments have been commercially available since the early 1940's. Since about 1955 diffraction gratings have also been used commercially as dispersing elements. What do you think their advantages are over prisms? We compare merits later in this Part.

Since about 1960 a totally different method of obtaining an infrared spectrum has increasingly found favour. This new method is based on an old idea of interference of radiation between two beams to yield an interferogram. An interferogram is a signal as a function of change of path-length between the two beams. You are aware that a spectrum is a signal as a function of frequency. The two domains of distance and frequency are interconvertible by the mathematical method of Fourier transformation.

In recent years computers have increasingly been interfaced to both dispersive spectrometers and Fourier transform spectrometers.

We therefore have three aspects of infrared spectroscopy which will be treated separately: dispersive spectroscopy, Fourier transform spectroscopy and computers.

2.1. INSTRUMENTATION FOR DISPERSIVE INFRARED SPECTROSCOPY

Infrared spectroscopy has passed through several interesting phases following the initial realisation of its commercial applications. This occurred in the 1940's particularly for the analysis of petroleum fractions followed by more general applications to analysis in chemical and manufacturing industries and elsewhere. During the 1950's a range of commercial instruments were marketed and many are still in service. It is therefore worth picking up the story with this generation of instruments.

2.1.1. Dispersive Elements – Prisms or Gratings?

The most popular prism material employed in spectrometers designed for routine work is NaCl, which is transparent to infrared radiation throughout the range 650–4000 cm^{-1}. You will recall that the dispersion of a prism depends on the change in its refractive index as the frequency of radiation changes. LiF has more favourable dispersion properties than NaCl at high wavenumbers, but is not transparent below about 1000 cm^{-1}. *It cuts off at 1000 cm^{-1}*. In order to extend the operating range below 650 cm^{-1}, spectrometers were designed which employed KBr and CsI. These materials are transparent to 400 and 200 cm^{-1}, respectively.

The popularity of prism instruments fell away during the late 1960s, when the improved technology of grating construction enabled cheap good quality gratings to be manufactured. Instruments

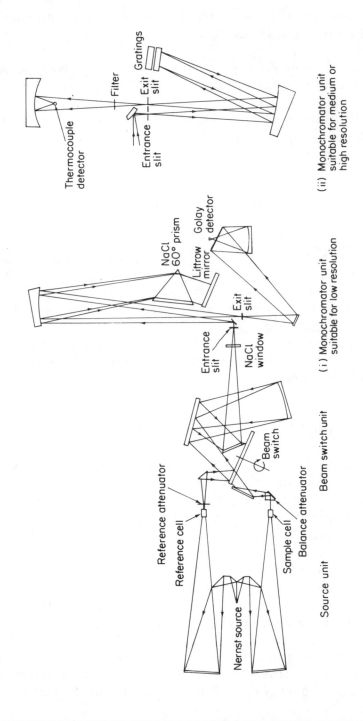

Fig. 2.1a. *The optical path of a double-beam infrared spectrometer with (i) a prism monochromator and (ii) a grating monochromator*

employing gratings as the dispersive element have the advantage of superior resolution, often over a wide frequency range. Prisms can, however, have the advantage of passing more energy than a grating.

2.1.2. Monochromators

The dispersive element is contained within a suitable monochromator. Fig. 2.1a shows the optical path of an infrared spectrometer using either (*i*) a prism monochromator or (*ii*) a grating monochromator.

As we saw in Part 1 dispersion occurs when energy falling on the entrance slit is collimated onto the dispersive element. The dispersed radiation is then reflected back to the exit slit beyond which lies the detector. The dispersed spectrum is scanned across the exit slit by rotating a suitable component within the monochromator. The rotation of this component may be linked to the chart drive mechanism and calibrated in wavenumbers. The width of the entrance and exit slit may be varied and programmed to compensate for any variation of the source energy with wavenumber. In the absence of a sample the detector then receives radiation of approximately constant energy as the spectrum is scanned.

SAQ 2.1a	Complete the following table with ticks for compatible combinations and crosses for incompatible combinations.

	Constant Resolution	Constant Energy
Constant Slit		
Programmed Slit		

2.1.3. Single Beam or Double Beam Instruments?

In the first generation of instruments the energy falling on the entrance slit originated from one beam of radiation from a continuous infrared source. A spectrum of the source background could be recorded in the absence of a sample and recorded again with the sample in the beam. The ratio yields the transmittance spectrum of the sample.

The single beam spectrum of the background is of course the spectrum of atmospheric absorption by CO_2 and H_2O in the instrument beam. Such a spectrum is shown in Fig. 2.1b and when you have examined it you will not be surprised that alternative instrumental arrangements have been devised so that the background spectrum does not exhibit CO_2 and H_2O absorption bands.

Alternatively, as shown in Fig. 2.1a, it is possible to have a double beam optical arrangement in which radiation from a suitable source is divided into two beams, a sample beam and a reference beam. These beams pass through a sample and reference path of the

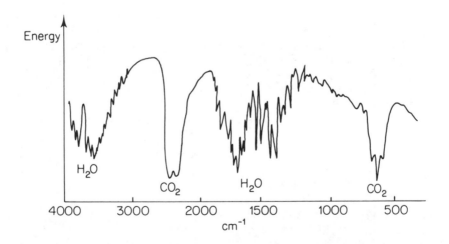

Fig. 2.1b. *A single beam infrared spectrum showing background atmospheric absorption*

sample compartment. Beyond the sample compartment but before the monochromator, these two beams become combined in space but separated in time by a rotating sector mirror acting as a beam switch. The sector mirror either reflects the sample beam or transmits the reference beam, via a system of mirrors, onto the monochromator which passes to the detector an alternating signal of the sample and reference beam at the frequency of the beam switch.

SAQ 2.1b

(*i*) Sketch a single beam infrared spectrum of any sample for which you can recall the general appearance of the spectrum. What about CO?

(*ii*) Sketch a double-beam infrared spectrum of the same sample.

2.1.4.   Reference Beam Attenuation or Ratio Recording?

In the monochromator which is shown in Fig. 2.1a, an attenuator is incorporated in each of the two beams. These are simple comb or wedge-type devices which move in and out with linear change in energy passing into the monochromator. The attenuator in the sample beam is a simple base-line adjustment. The attenuator in the reference beam is moved in and out of the beam according to the difference in energy between the two beams by the self-balancing servo, or null-balancing principle. That is, any imbalance between the sample and reference beams changes the signal falling on the detector from a continuous to a square wave function. The peak to trough amplitude of this wave is used to drive the attenuator back into balance and the movement is linked either mechanically or electrically to the pen of the spectrum recorder to measure transmittance of the sample at the particular wavenumber falling on the exit slit. In this system the mechanical movement of an attenuator limits the accuracy with which the measurement of transmittance or absorbance may be made.

Spectrometers with self-balancing servo systems were marketed up to the mid 1970's when improvements in electronics led to the introduction of ratio recording spectrometers which do not have the sources of error associated with the movement of a mechanical attenuator. These errors normally restrict the most accurate region of the transmittance scale to between 10 and 90%. In other words the number of absorbing molecules in the beam has to be adjusted by altering the path-length or concentration so that the sample is neither too weakly absorbing nor too strongly absorbing.

Ratio recording spectrometers measure the relative energy of the two beams in terms of electronic signals which are very high in signal to noise. This was made possible by developments of solid state electronics during the 1970's. These very low noise levels permit computer enhancement of very weak signals which would previously have been lost in the noise. For example Fig. 2.1c shows the spectrum of a trace quantity of an extracted contaminant from the surface of an electrical component.

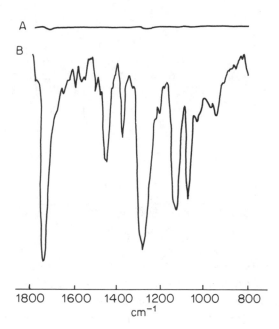

Fig. 2.1c. *Infrared spectrum of an evaporated residue (approximately 5 μg): A – normal spectrum, B – computer enhanced spectrum*

The original data (A), recorded from 5 μg material deposited on a KBr window, gives a maximum peak intensity of 0.75%T. The enhanced spectrum (B) follows a 200 times computer enhancement of the ordinate scale.

As well as enhancing spectra from samples with very high transmittance (or low absorbance) ratio recording spectrometers can lead to improvement for samples with very low transmittance (high absorbance). Fig. 2.1d shows the spectrum of a substituted aromatic hydrocarbon adsorbed on carbon black, with the sample prepared as a KBr disc. Under normal conditions, the maximum transmittance of this sample is 0.5%T over the range 3700–2700 cm^{-1} (A). Using a ratio recording spectrometer followed by computer enhancement clearly defined bands from the hydrocarbon are observed leading to assignment of OH, aromatic C—H and aliphatic C—H (B).

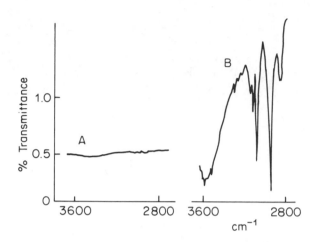

Fig. 2.1d. *Infrared spectra of a substituted aromatic hydrocarbon adsorbed on carbon black A – normal transmission spectrum with instrumental expansion B – computer replot from 64 accumalated scans with scale expansion and smoothing*

Further examples of computer manipulations of spectra will be discussed in Section 2.3.

SAQ 2.1c

Complete the following table by entering ticks for compatible combinations and crosses otherwise.

| | Dynamic Range | | |
	10–99%T	1–99%T	0.1–99.9%T
Ft-ir	√	√	√
Ratio Recording			
Servo Mechanism			

2.1.5. Sources and Detectors

The source and detector are the beginning and end of the optical path of an infrared spectrometer (Fig. 2.1a). They have a complementary role; a detector must have adequate sensitivity to the radiation arriving from the sample and monochromator over the entire spectral wavenumber region required. The sensitivity of the detector must also be adequate over the whole transmittance range required. To achieve this the source must be sufficiently intense over the wavenumber range and the transmittance range.

Sources of infrared emission have included the 'Globar' which is constructed of silicon carbide and which can be switched on and off electrically without undue thermal shock. The 'Nernst Filament' is a mixture of the oxides of zirconium, yttrium and erbium and has somewhat more power in the far-infrared than the Globar. The Nernst filament only conducts electricity at elevated temperatures and has to be heated before it strikes. It sometimes has to be either water cooled or air cooled and may crack under thermal shock. Recent sources include the 'Opperman' source which is self-igniting and self-regulating.

Most detectors have consisted of thermocouples with varying characteristics. For some purposes a 'Golay Cell' has been used. This operates on the principle of a large aperture beam heating a closed gas system which distorts a diaphragm which relays a secondary light signal to a photocell. The advantages may be outweighed by lack of reliability and high cost. Modern thermocouples have been developed with increased sensitivity and reliability. One limitation has been the small aperture and area of the sensitive area but this has been overcome by condensing the beam to a very small image by means of suitably coated caesium iodide lenses.

2.1.6. Amplification and Recording of Detector Signal

The detector in a double beam spectrometer is activated alternatively by sample and reference beam at the frequency of the beam switch. It is necessary to tune the amplifier receiving the signal from the detector to the frequency of beam switching to discriminate signals from the sample from signals arriving on the detector from other sources. In some instruments it is further possible to arrange an additional beam switch between the source and the sample. This enables the detector to discriminate radiation which passes from the source to the detector via the sample from, for example, radiation which is emitted by the sample. The weaker the signal from the sample beam the higher the electronic gain required within the amplifier is needed. Further signal improvement is possible using a longer amplifier time constant.

Although infrared spectra were initially plotted manually, the vast majority of infrared spectra have been recorded by an ink pen on chart paper. Typically pre-calibrated charts are used either from rolls or as single sheets. Increasingly spectra are stored on computer disc with retrieval and print out capacity.

2.1.7.   The Trading Rules

Weak signals are sometimes the result of attempting high resolution by reducing the slit width. It follows that high resolution has to be paid for in the currency of higher noise tolerance or longer time to obtain a spectrum.

If the desired result is very high signal to noise then both resolution and time, or possibly one or the other, must be sacrificed.

Finally, if speed is of the essence, as is frequently the case either from customer pressure or because of sample decomposition problems, then both resolution and signal to noise may have to be sacrificed.

Hence resolution, noise and time are three interdependent variables any two of which can be 'traded' for improvement in the third.

SAQ 2.1d

List common materials for prisms.

Comment on the relationship between relative atomic mass of the elements in the prism and the cut-off point to infrared radiation.

SAQ 2.1e

Sketch the simplest arrangement you can for obtaining a dispersed spectrum and scanning this across an exit slit.

SAQ 2.1e

SAQ 2.1f

In terms of the trading rules suggest ways in which the following may be achieved:

(*i*) High speed of recording a spectrum.

(*ii*) High speed of displaying a spectrum.

(*iii*) High resolution with no time restraints.

(*iv*) High resolution with severe time restraints.

(*v*) Low signal to noise with ability to rapidly record a spectrum.

SAQ 2.1f

2.2. INSTRUMENTATION FOR FOURIER TRANSFORM INFRARED (FT-IR) SPECTROSCOPY

This Section introduces some of the basic principles upon which Fourier transform infrared instruments are designed. You will be tested on all aspects covered at the end of the section where we have saved up several SAQ's! Do you know what is meant by apodisation? Do you know the advantages of Ft instruments? You will shortly.

'If God had intended us to carry out Fourier Transform Spectroscopy he would never have invented the rainbow.' In other words do we need to go beyond dispersive spectroscopy? The answer is – Yes, for the following reasons.

The essential problem of the prism or grating spectrometer lies with its monochromator. This contains narrow slits at the entrance and exit which limit the wavenumber range of the radiation reaching the detector to one resolution width.

∏ To illustrate the inefficiency of this method consider a dispersive spectrometer operating between 4000 and 400 cm^{-1} at a resolution of 1 cm^{-1}. At any time most of the radiation falls on either side of the jaws of the exit slit and only one wavenumber passes through. How much of the total radiation is being measured at any one time?

The answer is 1/3600.

If the resolution was changed to 10 cm^{-1} by increasing the width of the slit or to 0.1 cm^{-1} by closing down the slit how would the amount of energy change?

The answer is that it would be 1/360 or 1/36000. The higher the required resolution the lower the energy reaching the detector and, by the same trading rules as discussed in Section 2.1.8 the higher the noise and/or the longer the time of scan.

The circumstances of energy starvation are as follows.

Firstly, energy limiting factors may arise from samples which either absorb very highly, or reveal very little difference from the energy of the reference beam. In other words it is often the case that there is too much absorbing material either from the sample or from other components in the system or, at the other extreme, too little absorbing material because the sample is only present in trace amounts. These considerations lead to a limited dynamic range of absorption values for low sensitive spectrometers; typically instrumental transmittance values of between 90% and 10% of the energy in the reference beam which correspond to absorbance values of 0.08 and 1.00. This problem often occurs when special sampling accessories are used as will be discussed in Part 3. When high sensitivity dispersive spectrometers are used the dynamic range can be increased significantly as discussed in Section 2.1.4 up to the energy limitation inherent in the system.

Secondly, samples from which a very quick measurement is needed as, for example, in the eluant from a chromatography column, cannot be studied with instruments of low sensitivity since these cannot scan at speed.

Thirdly, when approaching the limiting resolving power of the instrument in order to achieve a high resolution spectrum, the spectral quality deteriorates if the energy reaching the detector becomes severely limited.

The fundamental limitations of energy by virtue of the use of slits in a dispersive spectrometer suggests this is an inefficient way to obtain a spectrum over a wide wavenumber range. Is there a better way?

An alternative approach was pioneered by Michelson and led to the methods of Fourier Transform Infrared Spectroscopy (Ft-ir).

2.2.1. The Michelson Interferometer

Towards the end of the 19th Century Michelson made many important optical discoveries in the U.S.A. one of which was the invention of the interferometer named after him. The optical layout is shown in Fig. 2.2a.

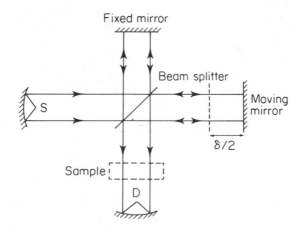

Fig. 2.2a. *Michelson interferometer (S = source, D = detector)*

A beam of radiation from the source, S, is focussed on a beam splitter which is constructed of material such that about half the beam is transmitted to a moving mirror which reflects the beam back to the beam splitter which then reflects part of this beam through a sample to a detector, D. The other half of the beam from the source is reflected from the beam splitter to a fixed mirror which reflects the beam through the beam splitter to the detector, D, via the sample. A suitable material with the necessary optical properties for beam splitting in the mid-infrared is KBr coated with germanium. In the far-infrared polyethylene terephthalate is used.

When the position of the moving mirror produces two beams

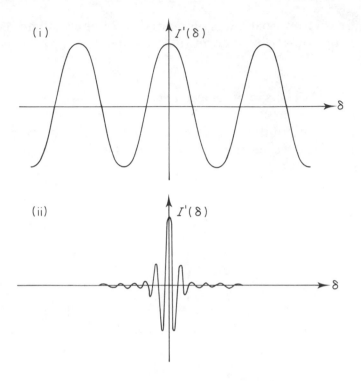

Fig. 2.2b. *Interferograms for (i) monochromatic radiation and (ii) white light*

travelling equal distances and falling on the detector, D, a strong signal should be obtained from the detector. When the moving mirror is scanned through a distance each side of the central position corresponding to zero path distance an interference pattern is registered by the detector. Michelson realised that this interference pattern contained spectral information. For example he deduced from the interference pattern of the red Balmer line in the spectrum of hydrogen that the line was a double line rather than a single line. The split nature of this line was subsequently confirmed by dispersive spectroscopy with improved resolution.

The development of the technique for recording an interference pattern by scanning the moving mirror through a distance of $\delta/2$ produces a total path difference of δ. The resultant interference pattern is shown in Fig. 2.2b for a source of monochromatic radiation (*i*) and for a source of polychromatic radiation (*ii*). The former is a simple cosine function but the latter is of more complicated form because it contains all the spectral information of the radiation falling on the detector.

2.2.2. Other Interferometers

The Michelson interferometer is the basis of many commercial instruments but there are a number of variants in which the interferogram is generated by creating a path difference other than by the optical arrangements shown in Fig. 2.2a. These will not be considered but they serve the same role in obtaining an interferogram from a source of radiation as the Michelson interferometer.

2.2.3. Fourier Transform Infrared Spectroscopy (FT-IR)

The mathematics of the conversion of an interference pattern into a spectrum is formidable. With characteristic genius Michelson constructed an early computer which he described in 1902 in the University of Chicago Press although there is no information on any of the results obtained.

The essential equations relate the intensity falling on the detector, $I(\delta)$, to the spectral power density at a particular wavenumber, $\bar{\nu}$, given by $B(\bar{\nu})$ are as follows:

$$I(\delta) = \int_{0}^{+\infty} B(\bar{\nu}) \cos 2\pi \,\bar{\nu}\delta.d\bar{\nu} \qquad (2.1)$$

which is one half of a cosine Fourier transform pair, the other being

$$B(\bar{\nu}) = \int_{-\infty}^{+\infty} I(\delta) \cos 2\pi \,\bar{\nu}\delta.d\delta \qquad (2.2)$$

These two equations are interconvertible. They are known as a Fourier Transform pair. The first shows the variation in power density as a function of difference in path-length which is an interferance pattern. The second shows variation of intensity as a function of difference in path length, which is a spectrum. Each can be converted into the other by the mathematical method of Fourier transformation. Early transformations were somewhat crude. The first numerical calculation of a spectrum from an interference pattern was by Fellgett in 1949. Interferometry was developed in the late 1950's notably at The National Physical Laboratories by Gebbie and his co-workers for the far-infrared where there were particular energy limitations for dispersive methods.

The essential experiment to obtain an Ft-ir spectrum is to obtain an interferogram with and without a sample in the beam and transforming the interferograms into spectra of (*a*) the source with sample absorptions and (*b*) the source without sample absorptions. The ratio of the former and the latter corresponds to a double beam dispersive spectrum.

The major advance towards routine use in the mid-infrared came with a new mathematical method or algorithm devised by Cooley and Tukey in 1965 for fast Fourier transformation. This was combined with advances in microcomputers which enabled these calculations to be carried out rapidly on-line rather than off-line.

2.2.4. Advantages of FT-IR

The method of fast Fourier transform infrared spectroscopy (Ft-ir) overcame the disadvantage of the measurement of one resolution element at a time and the disadvantage of the energy entering and leaving the monochromator being limited by narrow slits. These positive advantages of Ft-ir over dispersive ir are described respectively as follows:

'The Fellgett' or 'Multiplex Advantage'. Fellgett was the first to exploit the potential of interferometry for obtaining spectra from radiations of very low intensity such as those from astronomical sources. Since all the frequencies were recorded simultaneously rather than sequentially as in dispersive spectroscopy they were said to be 'Multiplexed'.

'The Jacquinot' or 'Throughput Advantage'. The cross-section of the beam of radiation in an interferometer can be much larger than that of a dispersive spectrometer although in many chemical and analytical applications the sample size limits the throughput to similar magnitudes irrespective of which instrument is used. When the sample has a large cross-section the Jacquinot advantage can lead to considerable improvements in signal to noise.

Further improvements resulted from longer collection time. The signal to noise increasing as the square root of the measurement time.

2.2.5. Sources and Detectors

In Section 1.1 we showed schematically the way in which a spectrometer could be constructed to obtain an absorption spectrum. A source of continuous radiation over the region of interest was used and we essentially collected two spectra and ratioed them to give the desired result. One was the spectrum of the radiation emitted by the source and the other the radiation of the source after it had passed through the sample. The ratio, of course, represents the absorption spectrum of the sample. Most Ft-ir spectra are also recorded in absorption mode and good emission characteristics are needed from the source over the ir region of interest. Sources are typically made from ceramic filaments operating at about 1500 K with temperature stabilisation.

What are the particular problems of Ft-ir detectors over others in the same region? Fairly obviously, response time has to be much more rapid than in dispersive spectroscopy. We are taking a large amount of energy quickly rather than a small amount of energy slowly. There are two commonly used detectors. The normal detector for routine work is a pyroelectric device incorporating deuterium triglycine sulphate in a temperature resistant alkali halide window. For extreme sensitivity mercury cadmium telluride has exceptional characteristics but has to be cooled to liquid nitrogen temperatures. This has a tendency to become saturated with energy and is best utilised for severe energy limited systems.

2.2.6.   The Moving Mirror

You may have guessed from consideration of the optical arrangement in Fig. 2.2a that the moving mirror is the most crucial component of the interferometer. It has to be accurately aligned and
capable of scanning between two distances so that the path difference corresponds to a known value. Careful readjustment to opti-
mise the alignment is often necessary. The effect of misalignment on
the background spectrum of the air path in the instrument is shown
in Fig. 2.2c.

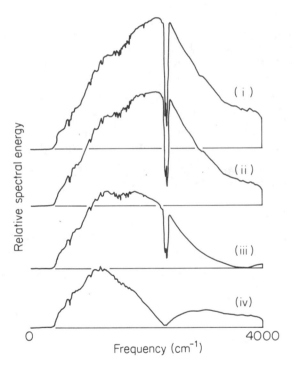

Fig. 2.2c.   *Single beam spectra taken when the interferometer is in
(i) good alignment, (ii) fairly good alignment, (iii) poor alignment
and (iv) very poor alignment*

The energy at high frequencies for (*ii*) is only slightly lower than for
(*i*) but in (*iii*) it is seen to be much lower. For very poorly aligned in-
terferometers, (*iv*) the phase may change so rapidly with frequency

that the phase correction routines which use a short double-sided portion of the interferogram may not be long enough to allow the true intensities to be calculated.

'Signal Averaging'. The main advantages of rapid scanning instruments is the ability to increase the signal to noise by signal averaging leading to increase of signal to noise proportional to the square root of the time. It follows that in a rapid scan interferometer the signal to noise, S/N, is related to the number of scans, by the relationship $S/N \propto n^{\frac{1}{2}}$.

∏   Let us consider a typical Ft-ir which scans from 4000 cm^{-1} to 400 cm^{-1} at a resolution of 1 cm^{-1}. This has a multiplex advantage over the equivalent dispersive spectrometer because all the information is measured at the same time rather than the information concerning the resolution intervals (1 cm^{-1}) falling on the exit slit sequentially. If it is assumed that noise is independent of signal then the 3600 cm^{-1} resolution intervals are equivalent to 3600 separate scans by the Ft-ir. What then is the improvement in S/N?

The answer is $(3600)^{\frac{1}{2}} = 60$

∏   Suppose we scanned in one second (a typical scan time) and signal averaged by repeat scanning 3600 times (taking 60 minutes rather than 1 second). What now is the further improvement in S/N?

The answer is a further $(3600)^{\frac{1}{2}} = 60$

You can see that there are diminishing returns for signal averaging in that it takes an increasingly longer time to achieve greater and greater improvement. Possibly you may be prepared to wait for 100 times the basic scan time for a 10-fold improvement but not 10,000 times the scan time for a 100-fold improvement.

The accumulation of a large number of repeat scans makes great demands on the instrument if it is to exactly reproduce the conditions. It is normal to incorporate a laser monochromatic source in the beam of the continuous source. The laser beam produces standard fringes by interference which can line-up successive scans accurately and can determine and control the displacement of the moving mirror at all times.

'Digitisation'. The interferogram is an analogue signal at the detector which has to be digitised in order that the Fourier transformation into a conventional spectrum may be carried out. The greater the number of data points the greater the correspondence to the true interferogram shape and the better the spectrum after transformation. Since we cannot have an infinite number of data points there is an optimum number for a given required resolution. If the spectral range lies between $\bar{\nu}_{max}$ and $\bar{\nu}_{min}$ the number of points required is

$$N = \frac{2(\bar{\nu}_{max} - \bar{\nu}_{min})}{\Delta\bar{\nu}} \qquad (2.3)$$

There are two particular sources of error in transforming the digitised information on the interferogram into a spectrum.

Firstly the transformation carried out in practice involves an integration stage over a finite displacement rather than over an infinite displacement. The consequence of this necessary approximation is that the apparent shape of a spectral line may be as shown in Fig. 2.2d in which the main band has a series of negative and positive side lobes or podes with diminishing amplitudes.

'Adopisation'. This is the removal of the side-lobes or podes by multiplying the interferogram by a suitable function before the Fourier transformation is carried out. The simplest method is called 'Boxcar Apodisation' which consists of truncating the interferogram in a manner which supresses the effect. A more effective way is to multiply the interferogram by a triangular function prior to Fourier transformation leading to a spectrum which has been subjected to 'triangular apodisation'. Many Ft-ir spectrometers offer operators the choice between these two apodisation options. When accurate band

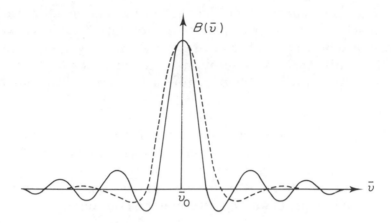

Fig. 2.2d. *Instrument line shape without apodisation*

shapes are needed more sophisticated mathematical functions may be needed.

A second source of error arises if the sample intervals are not exactly the same on each side of maxima corresponding to zero path difference.

'Phase Correction'. This is the correction which seeks to ensure that the sample intervals are the same each side of the first interval and which should correspond to $\delta = 0$.

'Resolution'. The resolution is limited by the maximum path difference between the two beams. It may be shown either by approximate theory or by more rigorous theory that the limiting resolution in wavenumbers (cm^{-1}) is the reciprocal of the path-length difference (cm).

∏　What path-length differences are necessary to achieve limiting resolutions of 1 cm^{-1}, 0.1 cm^{-1} and 0.01 cm^{-1}?

The answers are 1 cm, 10 cm and 100 cm.

You may feel that this simple calculation shows that it is easy to achieve high resolution. This is not the case since the precision of the optics and mirror movement mechanism become much more difficult to achieve at longer displacements of path-lengths. However, the technical difficulties of achieving this are usually less formidable than those of improving resolution to 0.1 cm^{-1} or 0.01 cm^{-1} by dispersive spectroscopy. Moving mirrors by up to 2 m has been achieved by Connes.

'Wavenumber Measurement'. The accuracy of wavenumber measurements is determined by the accuracy with which the displacement of the moving mirror may be measured. This accuracy is the same over the whole range of the instrument; this is in contrast to dispersive instruments where there may be considerable variation.

2.2.7. Computer Interface

We have noted that virtually all modern dispersive spectrometers are double beam thus automatically ratioing the background. Virtually all Ft-ir instruments are single beam and two spectra have to be recorded one without the sample, one with the sample, followed by computer subtraction of the first from the second. As described earlier the conversion of the interferogram into a spectrum by fast Fourier transformation also requires a good computer. Also the ability to accumulate a large number of repeat scans can only be done on a computer. It is clear that the development of Ft-ir has been made possible by the development of cheap, high speed, large memory capacity micro-computers which are on-line. Further aspects of computer applications will be considered in Section 2.3.

2.2.8. The FT-IR Spectrum

How then does an Ft-ir spectrum differ from a dispersive ir spectrum? The simple answer is, in many cases, not at all. In other cases it may be possible to obtain a much better spectrum by Ft-ir. Some typical examples are as follows.

Spectra were recorded in 30 seconds at a resolution of 4 cm^{-1}

Water is a notoriously bad solvent for use in infrared spectroscopy.

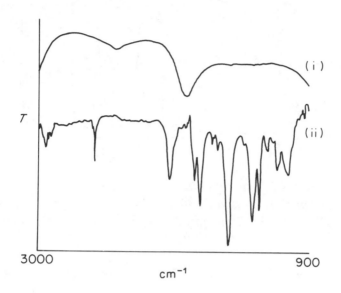

Fig. 2.2e. *The FT-IR spectrum of aspirin in water (i) 1% w/v (ii) 1% w/v with subtraction of the water spectrum.* © *Perkin–Elmer, 1984. Reproduced by permission of Perkin–Elmer Ltd.*

It absorbs very strongly and normal path-lengths are limited to up to about 25 μm. It also dissolves the common infrared cell materials (Part 3) such as NaCl, hence CaF_2 or B_aF_2 materials are needed.

Fig. 2.2e shows two spectra of a weak solution of aspirin in water. The potential problems posed by water are clear in the top spectrum (*i*). The power of a modern Ft spectrometer is evident in the lower spectrum (*ii*) which illustrates the result of subtraction of the water spectrum. Note band at 2350 cm^{-1} from dissolved CO_2.

A further area of application of Ft-ir is for examining solid polymeric samples of an intractable nature. Thus a copolymer of phenylene oxide and styrene is high melting, insoluble and hard. If rubbed with emery paper the powder deposited may be studied by diffuse reflectance methods to give a high quality spectrum as shown in Fig. 2.2f.

The ability to obtain spectra from trace amounts of materials is illustrated in Fig. 2.2g. The samples have been deposited on a micro

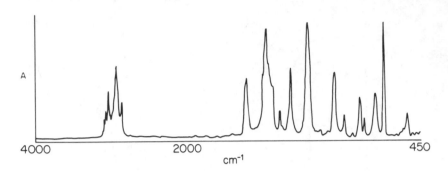

Fig. 2.2f. *Ft-ir diffuse reflectance spectrum of phenylene oxide-styrene copolymer recorded in 30 seconds at a resolution of 4 cm^{-1}. © Perkin–Elmer, 1984. Reproduced by permission of Perkin–Elmer Ltd.*

MIR crystal. We will explain MIR (multiple internal reflection) in Part 3.

There are many other cases where adequate spectra can only be taken by Ft-ir. Can you list suitable examples? Can you think of examples where dispersive spectroscopy is superior? In terms of the fundamental principles of the two kinds of instruments, it is likely

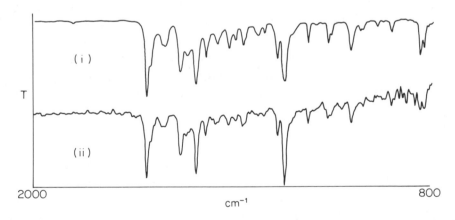

Fig. 2.2g. *The ft-ir spectra of phenacetin deposition on a micro MIR crystal (i) 1 µg sample; spectrum recorded in 30 seconds at 4 cm^{-1} resolution (ii) 50 ng sample; spectrum recorded in 2 minutes at 4 cm^{-1} resolution. © Perkin–Elmer, 1984. Reproduced by permission of Perkin–Elmer Ltd.*

that if it was necessary to repeat-scan one band as the state of the sample is changed, this may be done more efficiently on a dispersive spectrometer. This follows from the fact that an Ft-ir spectrometer measures the whole spectrum at a time whereas a dispersive spectrometer only measures one resolution element at a time.

SAQ 2.2a	Compare the advantages and disadvantages of Ft and dispersive infrared spectroscopy by completion of the table below. I have started you off by indicating that Ft-ir has the advantage of speed.	

	Dispersive ir	Ft-ir
Cost		
Resolution		
Speed		√
Computer access		
Signal/Noise		
Routine samples		
Intractable samples		
Wide cm^{-1} regions		
Narrow cm^{-1} regions		
Precision in cm^{-1}		
Accuracy in cm^{-1}		
Low stray light		
Double beam work		

SAQ 2.2b

Radiant energy is measured as a function of wavenumber, $\bar{\nu}$ by using a dispersive spectrometer.

Radiant energy is measured as a function of difference in path length, δ, using an Ft-ir spectrometer.

(*i*)   What is the name given to the pair of equations linking the two spectra resulting from these measurements?

(*ii*) What limits the wavenumber range of a spectrometer?

(*iii*) What limits the useful maximum δ that is employed in Ft-ir spectrometers?

SAQ 2.2c

(*i*) Sketch the optical arrangement for a Michelson interferometer.

(*ii*) Sketch an interferogram for a monochromatic source.

(*iii*) Sketch an interferogram for a polychromatic source.

(*iv*) In order to take an absorption spectrum of a sample where should the sample be located in the beam?

SAQ 2.2d

Complete the table below, by inserting two ticks appropriate to matching together the pairs of advantages.

	Jacquinot Advantage	Fellgett Advantage
Multiplex advantage		
Throughput advantage		

SAQ 2.2e   Briefly define the following terms in relation to Ft-ir:

Apodisation

Digitisation

Phase Correction

Signal Averaging

SAQ 2.2f   On lifting the lid of a Fourier transform spectrometer two sources of radiation were observed – a ceramic filament and a He/Ne laser. What are their functions?

SAQ 2.2f

SAQ 2.2g   An Ft-ir spectrometer is used to record a single beam spectrum from a single scan and a difference in path-length, δ, of 100 mm. A noise level of 1% is recorded, the S/N being 100/1.

(*i*)   How many scans are needed to improve the noise to 0.5% and to 0.01%?

(*ii*) What is the limiting resolution, in units of cm^{-1}?

(*iii*) How could a limiting resolution of 0.02 cm^{-1} be achieved?

SAQ 2.2g

2.3. COMPUTERS AND INFRARED SPECTROSCOPY

Many of us have lived through an explosion of methods of handling information brought about by developments in microelectronics and particularly the silicon chip. Can you believe that in the early 1960's there were advocates of hand calculating machines in which you turned a handle until a bell rang to tell you to move along and work out the next decimal place! The work on a computer which was so large that it filled a big hall in the early 1960's could be carried out on a small desk top microcomputer in the early 1980's. The purpose of this backward look is to emphasise that few fields move as swiftly and it will be fascinating to learn what later decades have to offer.

The new area associated with computers and communications, sometimes called information technology, has particular relevance to molecular spectroscopy. Consider the number of compounds for which molecular spectra have been or could be measured. Consider the possible detail in the infrared spectrum for each compound.

∏   Let us assume that there is spectral data for 10^6 compounds
       to be included in a data bank. If their infrared spectra are
       recorded over an interval of 3000 cm^{-1} at intervals of 1 cm^{-1}
       how much information is this per compound?

       Given a wavenumber and intensity value for each interval
       there are 6000 pieces of information for each compound.

How was information of this type gathered and stored as the subject
of molecular spectroscopy developed?

Firstly, by manual or photographic measurement.

Secondly, on the chart of a pen recorder.

Thirdly, within a computer memory store.

Can you see that the sheer volume of potential information is vast?
The problems of storage and retrieval by traditional methods are
daunting. The potential for holding this information in permanent
memory and storing on tape or disc for retrieval and processing is
a considerable challenge.

The first labour saving exercise is to interface the microcomputer
directly to the spectrometer. This enables the spectrum to be collected automatically and stored on disc or tape in digital form for
later retrieval. A normal copy on a chart presented in a conventional form can be taken simultaneously or at any future time by
printing a 'hard copy'.

Before considering some applications of computer interfacing it is
worth briefly looking at some of the more important features of
microcomputers which are relevant to interfacing them with spectrometers.

2.3.1. Microcomputers and Spectrometers

In this Section we will limit our discussion to the microcomputers interfaced to, or an integral part of a spectrometer. Let us start by looking at the essential feature of such microcomputers.

All computers consist of the four following fundamental components:

(*a*) An *input device* which in our case may be a spectrometer, or a keyboard, from which data is converted by a suitable interface into digital information in binary code for computer access.

(*b*) A *memory* in which information may be saved, stored and retrieved. The memory is in several forms:

— There is permanent memory (read-only memory or ROM). This is the repository of the various permanent programs or operating instructions, such as operating systems and interpreters etc.

— There is the volatile memory from which information is lost when the power is switched off and which holds working data (random-access memory or RAM).

— There is back-up memory in which information usually in the form of magnetic disks on which programs and data can be permanently stored until erased.

(*c*) The *central processor unit* (CPU) which processes information held in memory according to the programs.

(*d*) The *output devices* which are:

— Screens on which spectra can be presented in whatever form is clearest.

— Chart recorders for output of analogue information.

— Printers for output of digitised information.

It is necessary to know about the capacity of a computer or micro-computer you may use.

Why is this? So called 8-bit microprocessors are often equipped with sixteen parallel conductors (the address bus), used to address locations in the computer's memory. Each conductor represents one digit of a sixteen-bit binary number, therefore, the maximum number of memory locations which the microprocessor may address or access directly is $2^{16} = 65,536$. Each memory location is capable of storing 8-bits (or one byte) of information.

About half of this memory may be needed for the program and the CPU instructions leaving typically 32,000 bytes or 32K as working memory for data storage.

Microcomputers with higher capacity have been developed and these are based on 16 or 32 bit systems. They allow working memory capacities of typically 2,000,000 bytes or 2 Mbytes and higher.

SAQ 2.3a

For any computer with which you may be familiar estimate or look up the following

Memory Size Bytes K

Read only ROM

Random Access RAM

Backup

SAQ 2.3a

Computer operations are applicable to all forms of spectroscopy and have been well developed for dispersive and Ft infrared spectroscopy. The large number of examples outlined below serves to illustrate the range of these operations.

(*a*) Control of the instrument, eg, set scan speeds, scanning limits, optical conditions etc, and carry out predetermined sequences of measurements, possibly with automatic sample changing equipment.

(*b*) Save spectra by reading into the computer as the spectrum is recorded and saving into permanent memory. The digitised spectra may be stored on a floppy disc of small memory or on a Winchester disc of large memory.

(*c*) Load spectra from store into memory, display on monitor, make any changes needed in presentation and print out a 'hard copy' of the screen content on a chart or printer.

(*d*) Enhance weak bands by computer multiplication of weak signals against weak background spectra.

(*e*) Enhance overlapped bands such as sample bands which are otherwise obliterated against strong background spectra.

(*f*) Smooth random noise.

(*g*) Accumulate or Average weak signals by scanning the spectrum many times. The non-random signal becomes steadily stronger but random noise averages to zero.

(*h*) Subtraction of unwanted signals such as solvent spectra from solution spectra or known components of mixtures containing other components which require study in more detail.

(*i*) Format change such as transmittance to absorbance or *vice versa* or change in wavenumber region or change in the absorbance range.

(*j*) Peak Height measurements by selecting suitable base line points on each side of the band.

(*k*) Peak Area measurements as an alternative to peak height measurements for greater accuracy in certain applications.

(*l*) Derivative plots for enhancement of inflections on bands to provide improved resolution of shoulders due to overlapping peaks.

(*m*) Search within reference libraries held in computer store for identification of unknown spectra by selecting one or more of the best matches between unknown and known(s).

(*n*) Quantify the amount of particular components from peak height or area measurements in terms of spectra of known reference compounds.

(*o*) Print qualitative and quantitative information in suitable report form for one or more problems. In this connection microcomputers, or Data Stations, interfaced to spectrometers, may be programmed to carry out calculations in Basic, Fortran, Pascal, C-Language or others and may also be used as word-processors. Spectra or other forms of diagram may be incorporated in the text.

SAQ 2.3b | List the four main components of a computer.

SAQ 2.3c

A double sided floppy disc holds 80 K on each side, and a Winchester disc holds 25000 K of memory. Assuming an infrared spectrum requires 7 K at 1 data point per wave number between 600 and 3600 cm^{-1},

(*i*) How many spectra can be stored on the Winchester disc?

(*ii*) How many floppy discs would you buy to store 500 spectra?

(*iii*) If spectra were run between 2500 and 3500 cm^{-1} at 0.05 data points per wavenumber how many spectra could now be stored on a floppy disc?

The following examples, used elsewhere in this Unit, illustrate one or more of the 15 computer operations outlined previously.

Fig. 2.1c shows a weak spectrum of a residue of an ester in which we *enhance weak* bands against weak background absorption and present a strong spectrum by computer multiplication.

Fig. 2.1d shows the spectrum of a residue of hydrocarbon in which we *enhance overlapped* bands arising from strong background absorption using 64 *accumulated* scans and in which we also *smooth* random noise and *subtract* the background. Thus we have used four of the fifteen operations listed.

Fig. 2.2e shows the spectrum of aspirin as a 1% aqueous solution followed by *subtraction* of the water spectrum and in which we *enhance weak* bands.

Fig. 3.1c shows a spectrum in which we have changed the *format* of the ordinate scale to absorbance from that of % transmittance as shown in Fig. 3.1b.

In Part 6 you will be asked to determine the concentration of propan-2-ol in an unknown mixture (SAQ 6.1g). Spectra of the unknown and standard solutions are provided. If these spectra had been stored in a microcomputer the quantification could have been carried out very rapidly, with the aid of appropriate software.

Part 7 of the Unit contains a range of infrared spectral problems which will help to develop your skills at identifying unknown substances. We can use computers to help us on such tasks, by storing a library of spectra of known substances and then searching for a match with our unknown. As we have seen, however, such libraries require large memory capacity which may require each laboratory micro to be net-worked to a more powerful system.

2.3.2. A Glimpse into the Future

It will be apparent that I, as author of this Section could be in the laboratory at my place of work, using a Data Station belonging to my

Institution for producing this text rather than sitting at home using my son's BBC home computer! With more facilities the working, home and other environments could be net-worked and the concept that he or she *has gone to work* changes. The nature of the work will also change.

In my crystal ball I see a microcomputer which not only carries out the 15 operations described but many more including word-processing with the ability to incorporate calculations, diagrams and spectra from instruments and libraries. I also see the operator using the microcomputer at his home with telecommunication links to many laboratories and many instruments and libraries.

Let us consider the *worst case scenario* or possibly the *biggest challenge scenario* depending on whether you are a pessimist or an optimist. A large number of samples are collected, some of which are taken from an industrial complex for a whole range of quality control measurements and some are taken from the environmental surroundings where pollution is suspected.

If the analyses were to be attacked by infrared we are not far from having certain procedures of the kind embracing all or most of the 15 computer operations listed above. If so it would be feasible to load many samples (possibly hundreds) into an automatic sampling systems which may be pre-programmed to record spectra and, after suitable data processing operations, print out identities and amounts of various components associated with quality control and environmental contamination.

Obviously the computer output is no better than the input in terms of sample collection, quality of spectra and availability of software. For both qualitative and quantitative information, calibration data from known compounds is used. As new data is acquired this is automatically included into the data bank and the system becomes self-improving as it gains and learns more information with which to deliver answers to problems which are posed. Systems of this kind are known as 'expert systems'.

The prospect of this take-over of their role by a computer may alarm some operators into fears of redundancy or being replaced as a human operator by a computer or robot operator. Let us take an optimistic view. The challenge offered by the methods outlined are enormous in that measurements, previously inaccessible, may be made readily and for very large numbers of samples. This possibility poses considerable technical and scientific problems for which future workers will require greater skills of all kinds than their predecessors.

For convenience we have taken infrared in isolation but much work is in progress in interfacing techniques such as Chromatography–Spectroscopy–Computing. The strengths of chromatography lie in powers of separation of mixtures. The strength of spectroscopy is in identification of these separated components. Both chromatography and spectroscopy enable quantitative measurements to be made. With luck overlaps of selected bands will not occur in both chromatography and spectroscopy for particular quantitative measurements. The strength in computing is in handling and processing data.

Let us assume we require results of a very large number of analyses which are based on infrared (or some other spectroscopic method, such as mass spectrometry which lends itself to microcomputing interfacing on a number of grounds) gas chromatography (or some other separation method) or some combination of separation, identification and quantitative estimation. The opportunities for collecting much more data in the future raises a further question. What does it mean? Within the software of future computers will be statistical programmes for making better sampling decisions, better design of experiments and operating conditions, better interpretation of the results, better pattern recognition procedures. This area of work is known as Chemometrics.

The human tasks will be formidable but will put a higher premium on expertise. There may be no fewer in the labour-force but much more data to improve performance and conditions once it has been computer processed for ready understanding. Does this challenge grab you? If so Distance and Open Learning modes can span the gap between the end of formal education and the state of the art at the time which matters.

SAQ 2.3d   List twelve computer operations in decreasing
            importance (in your judgement and experience)
            for interfacing purposes with infrared spectrom-
            eters.

SAQ 2.3e   Give two-line definitions of the following com-
            puter jargon.

            Bits

            Floppy discs

            Bytes $\longrightarrow$

SAQ 2.3e (cont.)

Winchester discs

Minicomputer

Basic

Real time

Microprocessor

Robotics

Hard copy

RAM

C-language

Microcomputer

Expert systems

User friendly

Binary notation

ROM

Silicon chip

Word processor

Information Technology

SAQ 2.3e

Summary

Three aspects of the instrumentation used in infrared spectroscopy are dealt with in this Part of the Unit.

The first introduces those aspects of modern dispersive infrared spectrometers which are necessary for understanding the principles of operation and the capabilities and limitations of typical instrumentation.

The second treats Fourier transform spectrometers in a similar manner and also develops an understanding of the fundamental differences between dispersive and Fourier transform instrumentation, including differences in performance.

The final section introduces the essential features of microcomputers interfaced to, or an integral part of, a spectrometer. The advantages of such systems are illustrated by reference to the large number of useful functions which can be carried out.

Objectives

You should now be able to:

- compare the merits of single beam and double beam infrared spectrometers;

- appreciate the optical arrangement in double beam infrared spectrometers;

- use an infrared spectrometer more effectively through a knowledge of operating principles and optical components;

- understand how an interferogram results from the optical arrangement of a Michelson interferometer;

- appreciate the advantages of dispersive and of Fourier transform infrared spectrometers;

- appreciate the main features of a microcomputer and the many functions that are available when a spectrometer and computer are interfaced together.

3.   Sample Handling

Introduction

Infrared spectroscopy is somewhat unusual compared with most other analytical techniques, in that it is relatively easy to obtain spectra from solids, liquids, gases and polymer films and indeed with more specialised techniques from more complex samples like surface coatings and emulsions.

You have already studied the principles governing the absorption of infrared radiation and the instrumentation required to detect this in Parts One and Two of this Unit. In this Part we shall study how samples can be introduced to the instrument, the equipment required and any pretreatment necessary for the different states of matter mentioned above. This should enable you to place the theoretical knowledge gained so far in a laboratory context.

In these early sections of this Part of the Unit we shall restrict ourselves to simple solids, liquids and gases where we have at least 10 mg of sample, I hope you have access to an ir instrument so you can try the techniques below for yourself. In later sections we shall look at more specialised techniques for surface films, polymers, samples from chromatographic techniques, and at variable temperature work and techniques for handling small samples.

We have not put many SAQs in this Part of the Unit, but your own questions and problems will arise when you first start to prepare samples. If you have studied this Part and have it with you in the laboratory, you should get on very well.

A choice of sampling technique exists for all states of matter. The one chosen depends on the application. The techniques we shall be studying in detail are as follows.

Techniques for Solids

(*a*) a solution in a suitable solvent,

(*b*) a suspension in a liquid (mull),

(*c*) mixed with an alkali halide and pressed into a disc,

(*d*) reflectance techniques.

Techniques for Liquids

(*a*) a solution in a suitable solvent,

(*b*) a thin film of pure liquid,

(*c*) a gas if the vapour pressure is high enough,

(*d*) reflectance techniques

Techniques for Gases

(*a*) a gas,

(*b*) a solution in a suitable solvent.

Techniques for Polymers

(*a*) a solution in a suitable solvent,

(*b*) a thin film,

(*c*) a thin film evaporated on an alkali halide plate,

(*d*) a thin slice cut from a larger sample,

(*e*) reflectance techniques.

This list is not exhaustive, but seems enough choice to be going on with!

Note the very wide range of techniques even for each sample type.

Before we start can you think of three reasons why this choice is important, I realise this will be difficult for you, but the answers to the questions you ask yourself here should appear in the text which follows.

Well it depends what you want from the spectrum, why you are recording it in the first place and how much time and money you are prepared to invest. If you were interested in whether the following reduction of an ester to an alcohol had gone to plan,

$$PhCOOCH_3 \xrightarrow{\quad LiAlH_4 \quad} PhCH_2OH$$

a liquid film spectrum of starting material and product would suffice. A check in the carbonyl region (1680–1830 cm^{-1}) and in the O—H stretching region (3100–3600 cm^{-1}) of the product, would quickly give you the required information (see Part Seven for detailed discussion). On the other hand if you were interested in hydrogen bonding in alcohols this technique would tell you nothing apart from the obvious, ie hydrogen bonding is present in the pure liquid. Here you would need to spend time preparing solutions in a non-polar solvent of accurately known concentration (see Part Five).

Molecular association in pure solids and liquids, and indeed solvent–solute interactions can influence the spectra obtained from polar molecules and these factors should always be borne in mind when the technique is selected for a particular sample. Specific examples will be given under individual techniques as they are discussed below.

It is unusual to record samples in the vapour state because of the difficulties of obtaining a satisfactory spectrum at low vapour pressures. The added complication of rotational fine structure and the presence of shifts (often quite large) in the frequency of absorption of polar groups has also to be kept in mind. The most easily reproduced spectra are obtained in dilute non-polar solutions. This, however, takes time and it is usual to resort to liquid films, KBr discs and mulls for routine samples.

3.1. A LITTLE HISTORY

Early infrared instruments recorded percent transmittance over a linear wavelength range. Since the early 1960's it became commonplace in more expensive instruments to have the choice of absorbance or transmittance as a measure of band intensity. It is now unusual to use wavelength for routine samples. Inverse wavelength units are used. This is a wavenumber scale and the units used are cm^{-1}. Wavenumbers are the reciprocal of the wavelengths expressed in cm. Wavenumber $\bar{\nu}$, wavelength λ and frequency ν are interrelated

$$\bar{\nu} = \frac{1}{\lambda} = \frac{\nu}{c} \qquad (1.3)$$

The output from the instrument is referred to as a spectrum and would perhaps logically have the form of Fig. 3.1a.

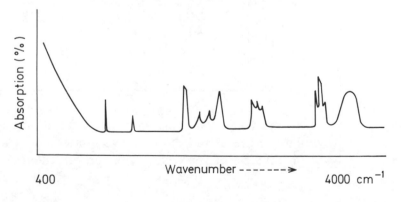

Fig. 3.1a. *A conventional 'graphical' presentation*

This would be the conventional method of drawing a graph, with ordinate increasing from bottom to top and abscissa increasing linearly from left to right. However, most commercial instruments present the spectrum with wavenumber decreasing from left to right. Older instruments may present the spectrum as linear in wavelength (hence non-linear in wavenumber).

Let's have a closer look at this. The question below should also reinforce some of the earlier material.

SAQ 3.1a

Complete the following table with the wavenumber corresponding to the wavelengths given (all wavelengths are in micrometres, 10^{-6}m). Express the wavenumbers in units of cm^{-1}.

Let me do the first one:

2.50×10^{-6} metre = 2.50×10^{-4} cm, and the reciprocal of this = 4000 cm^{-1}, now go ahead and complete the table.

Wavelength (μm)	Wavenumber (cm^{-1})
2.50	4000
3.33	
5.00	
10.00	
15.00	
20.00	
25.00	

Now draw a graph of wavenumber against wavelength in the form below, I have plotted the first and third point for you, is your graph linear?

$\longrightarrow$

**SAQ 3.1a
(cont.)**

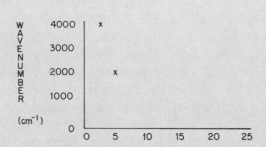

Now find a piece of paper and draw up a linear grid with the scale 4000–400 cm^{-1} along the abscissa, and write % Transmittance up the ordinate.

This used to be the normal way of presenting an infrared spectrum. Remember that few ir bands normally appear in the region 4000–1800 cm^{-1}, with many bands between 1800 and 400 cm^{-1}; do you see any problems with this type of spectrum presentation? How would you solve it?

The solution to the problem (the part of the spectrum rich in information takes up only about a third of the chart paper) which you hopefully recognised in the question above is to change the scale on the abscissa at 2000 cm^{-1}. This serves to crush up the region between 4000 and 2000 cm^{-1} and expand the region between 1800 and 650 or 400 cm^{-1}. This region after all seems to contain a lot of absorption bands. The ordinate scale may be presented in '% transmittance' with 100% at the top of the chart. An absorption band is therefore presented by the pen moving down the paper. The introduction of linked microcomputers allows the spectrum to be recalculated and presented as absorbance versus wavenumber. (This was always possible with more expensive instruments). An example of these two types is given in Fig. 3.1b and 3.1c for the same compound.

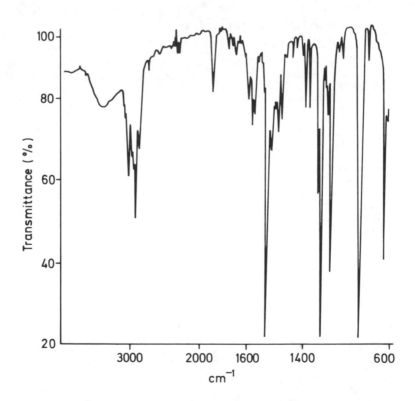

Fig. 3.1b.   *Transmittance v Wavenumber*

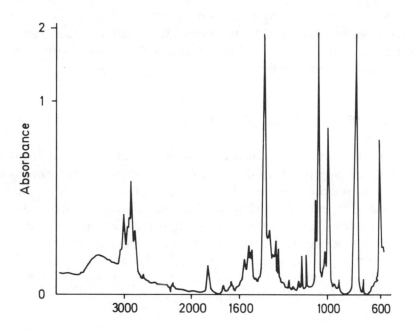

Fig. 3.1c. *Absorbance v Wavenumber*

Note the differences.

It almost comes down to personal preference which is better, but the % transmittance scale is now traditional for spectral interpretation purposes but unsuitable for quantitative work (see Part Six).

The older wavelength presentation also serves to crush the region between 4000 and 2000 cm^{-1}, corresponding to 2.5 to 5 μm. This type of presentation expands the region between 1000 and 600 cm^{-1}. The wavenumber range may vary depending on the instrument used and the application. The region normally studied, again traditionally, is 4000–650 cm^{-1}. This range is now commonly extended to 200 cm^{-1} for more expensive instruments.

Why is 4000–650 cm^{-1} chosen for study?

The early instruments used sodium chloride optics and NaCl does not transit below 650 cm^{-1}. With the introduction of diffraction gratings this range can be extended to 200 cm^{-1}, below which more severe instrumental problems occur. We shall also see later that the region of the infrared which contains the most information for organic chemists is 4000–650 cm^{-1}.

3.2. EXPERIMENTAL METHODS

The sample is contained in a *cell* having a path which is transparent to the incident infrared radiation. The sample is placed in this path. It is obvious that the parts of the cell in the beam must be constructed of materials that do not themselves absorb radiation. These 'windows' are normally alkali halides. The cheapest material is sodium chloride but other commonly used materials are potassium bromide, potassium chloride, caesium bromide and caesium iodide. These materials are almost transparent from 4000 cm^{-1} to a cut-off frequency which is given in Fig. 3.2a.

Compound	Cut-off/cm^{-1}
NaCl	650
KCl	400
KBr	320
CsBr	200
CsI	150

Fig. 3.2a. *Cut off frequencies for some alkali halides*

The choice of material will again depend on the application. Your sample may be in aqueous solution; sodium chloride would not then be a good choice! Here you would use calcium or barium fluoride (but watch the pH!) or silver chloride.

SAQ 3.2a

(*i*) Why do we not use sodium chloride plates for aqueous solutions?

(*ii*) Why must we watch the pH when using calcium or barium fluoride?

(*iii*) What is a suitable pH range for calcium and barium fluoride windows?

We shall now run through the mechanics of preparing the samples then look at some 'good' and 'bad' results in Section 3.3 and finally look at the range of cells available.

3.2.1. Liquid Films

The quickest method. Simply place one small drop of the liquid on a sodium chloride plate and place a second plate on top to make a sandwich. The liquid will spread out to form a thin film. This is then mounted in a cell-holder and placed in the front beam (the sample beam) of the instrument.

∏   This method is normally not used for volatile (b.p. < 100 °C) liquids. Why do you think this is?

Volatile liquids will evaporate in the heat of a beam. A sealed short path-length cell would then be used.

3.2.2. Solutions

The sample is dissolved in a suitable solvent (see below), a cell of known path-length is filled and placed in the spectrometer. The concentration is usually in excess of 0.01 mol dm^{-3}. This should be compared with the ultraviolet technique where concentrations one thousand times less are used routinely.

This statement tells us little about the amount of sample we need so let's look at an example.

∏   If we have a compound with a relative molecular mass of 200, and want a 5% w/v solution, how much sample do we realistically require? What is the molarity of this solution?

Well, it is difficult to make up less than 1.0 cm^3 of solution accurately, so let's assume we have a 1.0 cm^3 volumetric flask.

We need 5.0 g made up to 100 cm^3, so here we need 0.05 g, ie 50 mg.

The molarity of the solution is calculated as follows:

Concentration is 5 g dm^{-3}

The relative molar mass is 200,

so the molarity $= 5/200 = 0.025$ mol dm^{-3}

It is possible to use less sample by employing low volume pipettes of say 0.1 and 0.01 cm^3.

∏   What is a 'suitable' solvent? Note down *three* criteria you would consider important.

I can think of lots, I hope the three you have written down are in the list below.

— It has to dissolve the compound.

— It should be as non-polar as possible, to minimise solute-solvent interactions.

— It is important that it does not *react* with the compound.

— Most important, it should not strongly absorb infrared radiation. Other lesser considerations can also be important.

— It should be volatile and hence easily removed to allow retrieval of the sample, if valuable.

— It should also be pure and dry.

— It should not be viscous, since samples are contained in short path-length cells.

You may have thought of others.

No solvent exists which meets all these criteria. For example all solvents absorb to some extent in the region 4000–650 cm^{-1}.

SAQ 3.2b

Examine the spectra of four solvents, Fig. 3.2b–e (0.02 mm path-length).

Is it possible to use one of these solvents only and get a full-range spectrum without interfering peaks?

Is it possible to use two of these solvents separately and combine the information in the transparent regions? If so, which two solvents?

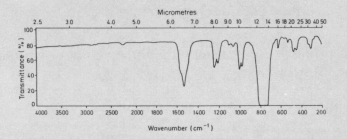

Fig. 3.2b. *Tetrachloromethane*

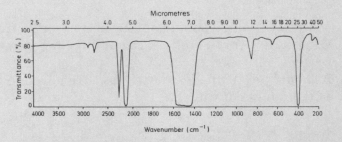

Fig. 3.2c. *Carbon disulphide*

→

**SAQ 3.2b
(cont.)**

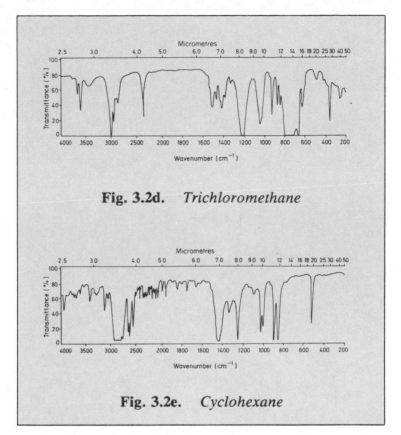

Fig. 3.2d. *Trichloromethane*

Fig. 3.2e. *Cyclohexane*

SAQ 3.2b

3.2.3.   Mulls

This is the least time-consuming method for obtaining spectra of solids. Normal crystals will scatter infrared radiation since they will generally be larger than the shortest wavelength used (2.5 × 10^{-6} m). The crystal size must therefore be reduced. This is normally done by grinding the solid (using an agate mortar and pestle), then suspending it (50 mg) in 1–2 drops of the mulling agent, followed by further grinding until a smooth paste is obtained. 'Tailing' spectra are produced when scattering is present. Again the ideal mulling agent would be transparent to infrared between 4000 and 600 cm^{-1} and again no such agent exists. Three mulling agents are commonly used, liquid paraffin sometimes referred to as Nujol, (an early trade name), Voltalef 3S oil (a halogenated hydrocarbon), and hexachlorobuta-1,3-diene. Their spectra are given in Fig. 3.2f, Fig. 3.2g and Fig. 3.2h below.

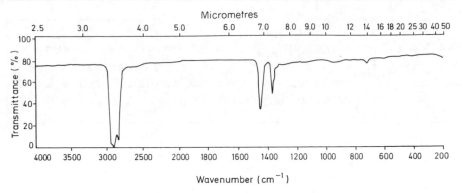

Fig. 3.2f. *Liquid paraffin*

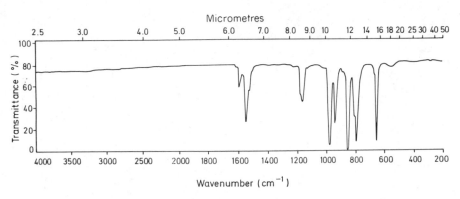

Fig. 3.2g. *Hexachlorobuta-1,3-diene*

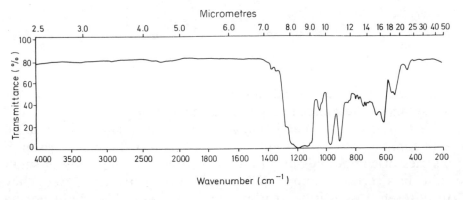

Fig. 3.2h. *Voltalef 3S Oil*

SAQ 3.2c

Write in the space below two possible combinations of mulling agents, which when used separately allow you to examine the infrared region between 4000 and 600 cm^{-1}.

This technique has the advantage of speed but suffers many disadvantages. The molecules will be associated leading to large band shifts (up to 20 cm^{-1}) and changes in intensity. Bands may also be split compared to dilute solution spectra. This makes comparison of a series of compounds uncertain. If a study is being made of small shifts in a series of compounds, say the C=O stretching frequency in differently substituted carboxylic acids, then this would not be the way to do it. The best method would be in dilute solution in a non-polar solvent.

3.2.4. Alkali Halide Discs

Here the solid (a few mg) is mixed with a suitable dry alkali halide (100–200 mg), ground in a mortar or ball mill and subjected to a pressure of about 10 ton/sq.in. in an evacuated die. This sinters the mixture and produces a clear transparent disc. Commercial equipment is available to produce the correct diameter (10–15 mm) to fit the spectrometer beam. Again it is essential that the crystal size is reduced to about 2.5 micrometres.

The most commonly used alkali halide is potassium bromide which is completely transparent in the commonly scanned region. The spectrum quality can often be improved by storing the KBr in an

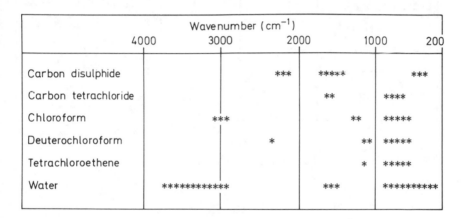

Fig. 3.2i. *Common solvents (path-length 2 mm)*

Fig. 3.2j. *Mulling agents (thin film)*

oven at a temperature above ambient and also by heating the resultant disc in a vacuum oven before recording the spectrum, both these measures reduce moisture content.

Fig. 3.2i, Fig. 3.2j and Fig. 3.2k summarise the regions in which the common solvents, mulling agents and alkali halides have a transmittance of greater than 25% (white) and less than 25% (starred) for the conditions specified. Note that I have used a linear wavenumber scale here, not the normal scale seen on spectra..

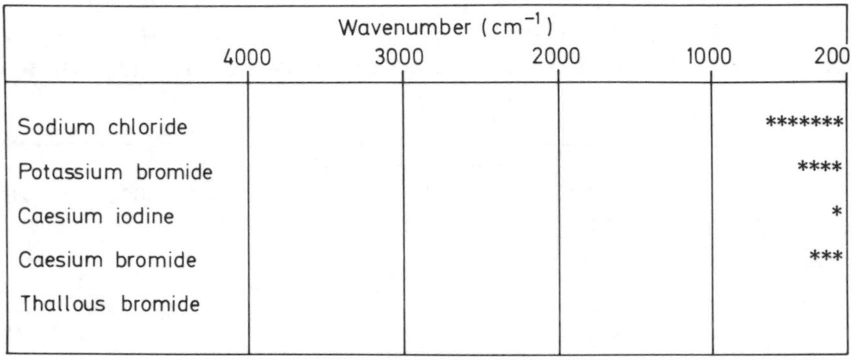

		Wavenumber (cm^{-1})			
	4000	3000	2000	1000	200
Sodium chloride					*******
Potassium bromide					****
Caesium iodine					*
Caesium bromide					***
Thallous bromide					

Fig. 3.2k.   *Alkali halides*

3.2.5. Gases

Special gas cells are marketed with path lengths from 10 cm to more than 20 m (see below). The cells have two taps and can be filled by flushing if the sample is plentiful or by attachment to a vacuum manipulation line. In the latter case the pressure in the cell can be controlled and monitored.

Gas spectra tended to be a specialist area until fairly recently. However, it is now more common to study samples from gas chromatographs and special micro cells have been developed for the purpose. With the advent of Fourier transform spectroscopy and computer control it is now possible to obtain a spectrum in a second or two, so that real-time spectra from GLC columns can be obtained. This technique is now also being applied to High Performance Liquid Chromatography samples.

Spectra obtained from compounds in the vapour phase differ considerably from those of the liquid or solid phase. Less molecular association is present so that large shifts to higher wavenumber (100–200 cm^{-1}) in polar groups should be expected (OH, C=O, C—O etc) when compared with spectra obtained from the liquid or solid state.

3.2.6. Spectral Techniques for Polymers

The available methods are:

(*a*) a solution in a suitable solvent,

(*b*) a thin film,

(*c*) a thin film evaporated on an alkali halide plate,

(*d*) a thin slice,

(*e*) reflectance techniques.

If the polymer is a thermoplastic (polyethylene, polystyrene) it can be softened by warming and pressed, in a hydraulic press, into a film thin enough for use. This can then be mounted on card (with an aperture) and the spectrum recorded. If this is not possible, eg perspex, the polymer can be dissolved in a volatile solvent and the solution allowed to evaporate on an alkali halide plate. This leads to a thin film suitable for study. Some polymers such as cross-linked synthetic rubbers can be microtomed, ie cut into thin slices with a blade. If the polymer is a surface coating reflectance techniques can be used, (see Section 3.5.2).

3.3. OPTIMISATION OF SIGNAL

The problems commonly encountered in obtaining good quality infrared spectra are usually due to either instrumental factors or arise from bad sample preparation. Part Two of this Unit should have helped you to get the best out of an infrared spectrometer and we shall now concentrate on sample preparation. There are quite a few pitfalls to avoid and you will soon be in a good position to avoid them.

Most of the techniques so far covered have used a cell with alkali halide windows or alkali halide plates. These materials are easily damaged and should only be handled by the edges. They should be kept in an oven when not in use. This is to prevent uptake of water. Samples are often wet and water will pit and fog the plates. Plates should therefore be regularly polished using kits which are available for the purpose. Calibration of fixed path-length cells is also important since the path-length will lengthen if wet solvents or samples are used.

3.3.1. Liquid Films

With a little care there is not a lot to go wrong here. If, however, the plates are pitted too much sample may be held in the film and the resulting spectrum may be too intense. An example is shown in (Fig. 3.3a).

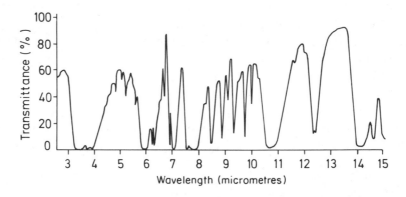

Fig. 3.3a. *A spectrum obtained when too much sample was contained between the plates*

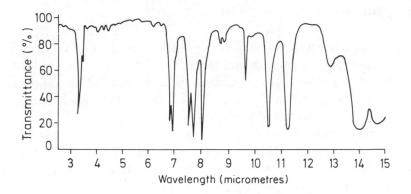

Fig. 3.3b. *A good quality spectrum of a volatile sample*

A common problem is due to sample volatility. When the spectrum of a volatile sample is recorded it becomes progressively weaker because evaporation takes place during the recording period. Fig. 3.3b shows a good quality spectrum obtained for a volatile sample – a sealed cell was used. Now examine Fig. 3.3c which shows a second spectrum of the sample. The sample has evaporated during the recording period. Liquids with boiling points below 100 °C should be recorded in solution or in a short path-length (0.01 mm) sealed cell.

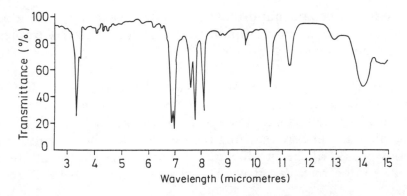

Fig. 3.3c. *A poor quality spectrum of a volatile sample evaporation of the sample has occurred during the recording*

3.3.2. Solutions

We have already established that solvents should be dry and pure, and that in order to obtain sample data over the range 4000–650 cm^{-1} two solvents will have to be used separately. The solution is placed in the sample beam and pure solvent in the reference beam of the instrument. Matched path-length cells are used. In a perfect world no absorption peaks from the solvent should be present. This will only be approximately true because solutions as concentrated as 10% w/v are sometimes used. The concentration of solvent in the reference beam will now be significantly greater than that in the sample beam.

∏   If the concentration of solvent is significantly greater in the reference beam what effect do you think this would have on the spectrum produced by the spectrometer.

Negative peaks may appear in the regions of solvent absorption, as you can see in Fig. 3.3d.

These effects will also appear if the path-lengths in the two beams are not identical. This is often caused by pitting of the cells or by water damage. Variable path-length cells are available; the path-length of such a cell can be varied after filling, and the best conditions obtained by alteration of the reference beam path-length. Both the effects above can therefore be diminished.

These effects become important in quantitative analysis as does the accurate measurement of path-length. Part Six deals with these points in more detail.

Can we believe our eyes and interpret peaks in regions of solvent absorption, eg in the region 650–820 cm^{-1} for a chloroform solution? Most unlikely, since absorption of radiation in both beams will be high, the energy reaching the detector will be very low and spurious peaks will be produced. Solutions at two concentrations may have to be recorded if weak peaks eg overtone bands, are of interest.

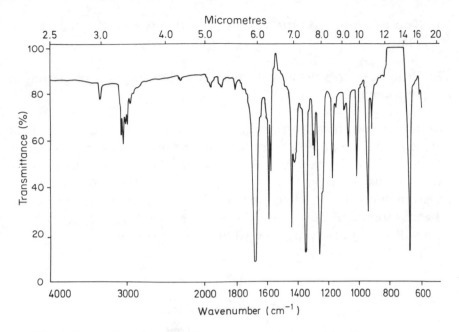

Fig. 3.3d. *The infrared spectrum of a sample dissolved in tetrachloromethane*

3.3.3. Mulls

A long list of potential problems exist and many are frequently encountered. We will now examine the most common ones.

(*a*) The ratio of the sample to mulling agent is wrong.

If the sample concentration is too low you get a spectrum of the mulling agent and no sign of the sample. On the other hand, if too much sample is used giving a very thick paste, the mull will not transmit radiation.

A rough guide to mull preparation is to use a micro spatula tip of solid to 2–3 drops of mulling agent. Aim to keep the most intense absorption band just on scale.

(*b*) Too much or too little mull is placed between the alkali halide plates.

Too little leads to a very weak spectrum showing only the strongest absorption bands.

Too much mull leads to poor transmission of radiation so that the base line may be at 50% transmittance or less. It is sometimes possible to reduce the energy of the reference (back) beam to a similar extent, by use of an attenuator. This will bring the baseline back to a reasonable value in the region of 90%. I hope you are able to use a beam attenuator next time you operate an infrared spectrometer. You will then fully appreciate what I am getting at.

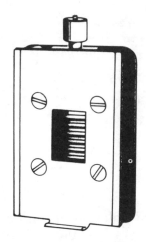

Fig. 3.3e.   *An attenuator*

Beam attenuators work on a venetian blind or moving comb principle. If you use one remember that you are skating on thin ice, since the energy reaching the detector from both beams has been cut and spurious peaks can result.

(*c*) The crystal size of the sample is too large.

This leads to scattering of radiation which gets worse at the high frequency end of the spectrum. The effect is dramatically illustrated by the spectrum in Fig. 3.3f. Additionally, bands are distorted and consequently their positions move leading to the possibility of wrong assignment. Since crystal hardness varies, no 'rules' apply but samples should be ground for at least one minute. In the case of very hard crystals a ball mill can be used to powder the sample prior to mulling. Some crystals, such as those of naphthalene, shear so that no matter how small the crystal size they can still present a flat surface to a beam of radiation. The reflection of infrared radiation leads to skewed bands.

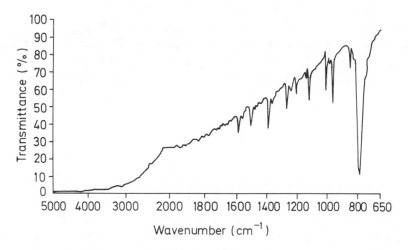

Fig. 3.3f. *Scattering of radiation in an infrared spectrum caused by the large crystal size of the sample*

(*d*) The mulling agent is difficult to remove.

This is usually unimportant except for valuable samples. Such samples should be dissolved in a pure volatile solvent.

(*e*) The mull does not cover the whole beam.

If the mull is not spread over the whole plate area, the beam of radiation passes part through the mull and part through only the halide plates and odd effects are produced. For example, strong bands will have flat ends and will not reach the bottom of the chart.

∏   Can you think of four criteria for obtaining a good quality
       spectrum from a mull?

The four most important are probably the following.

(*i*)   The most intense absorption bands due to the sample should
         be more intense than the bands from the mulling agent.

(*ii*)   Little or no scattering of radiation should occur.

(*iii*) No water should be present.

(*iv*)   The strongest bands due to the sample should almost fill the
         chart but should not peak at less than about 10% transmit-
         tance. (If they do they will lose their sharpness and the more
         so for the older optical-null double beam spectrometers.)

3.3.4.   Alkali Halide Discs

Molecular distortions induced during the high pressure preparation of discs cause band positions to be significantly different from
those observed in solution spectra. For the moment, however, we
are mainly concerned with the problems of preparing discs, and as
you will see they are very similar to those we have just considered
for mulls. I will take each point in turn.

(*a*) The ratio of the sample to alkali halide is wrong.

Surprisingly little sample is needed and if you use around 2 to 3 mg
of sample with about 200 mg of halide you should get good results.

(*b*) The disc is too thick or too thin.

Thin discs are fragile and difficult to handle and thick discs transmit
little radiation. A disc of about 1 cm diameter made from about 200
mg of material usually results in a satisfactory thickness of about 1
mm.

(*c*) The crystal size of the sample is too large.

Excessive scattering of radiation results and particularly so at high wavenumbers. You will recall Fig. 3.3f.

(*d*) Ionic samples can suffer double decomposition reactions and the alkali halide may be difficult to remove.

(*e*) The alkali halide is not perfectly dry.

This results in the appearance of bands due to water and difficulties in sample preparation. It is difficult to avoid but the alkali halide should be kept desiccated and preferably warm at the same time. KBr is widely used for making discs.

Generally speaking discs give better spectral results than do mulls. This is mainly due to sharper bands and hence better resolutions, and the presence of bands solely from the sample itself.

The spectrum resulting from a well prepared KBr disc is given in Fig. 3.3g. Check that the following three criteria for a good quality disc have been met.

(*i*)   Little scattering of radiation should occur.

(*ii*) Water should be absent.

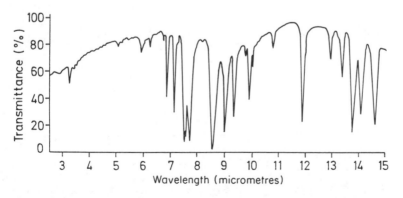

Fig. 3.3g.   *The infrared spectrum of a sample prepared as a KBr disc of good quality*

(*iii*) The disc should be of appropriate thickness (hence the strongest bands should fill the chart but not peak at less than about 10% transmittance).

SAQ 3.3a

Let's suppose you are running an infrared service and three samples await your attention. You can delegate two of the samples below to other workers. Since it is Monday morning and you are not quite awake yet, which would you choose to record? (All spectra are to be recorded using KBr discs).

(*i*)   a polycyclic hydrocarbon melting at 125 °C,

(*ii*) phenol,

(*iii*) trimethylamine hydrochloride.

There is not much to pick and choose between discs and mulls. They both suffer similar disadvantages. It is, however, usually quicker to prepare a mull and this is probably the reason for making it the more popular. Both need a finely ground sample and poor grinding is the reason for the majority of bad spectra. The mulling agents are much less likely to react with your sample and as you should have learned from the answer to SAQ 3.3a, ionic or very polar compounds do not usually give satisfactory discs.

3.3.5. Gases

Since spectra of gases reveal sharper bands than those from the condensed states, more information may be available from higher resolution studies. The main problem is leaking cells. Since most spectra are recorded with the pressure below one atmosphere, a frequent problem is air leaking into the cell. This is made more important since slow scan speeds are usually used to attain high resolution. The alkali halide windows used are larger, more expensive and need careful treatment.

3.3.6. Calibration of Spectra

We have a final problem. After all these careful preparations, band frequencies may be offset because of lack of frequency calibration. Since infrared spectroscopy is mainly used for structure determinations, it is important that frequencies are accurate and reproducible from week to week and from instrument to instrument. Methods have therefore been developed for the calibration of spectra.

The frequency calibration of a spectrometer may be inaccurately presented simply because the chart paper has been positioned badly. A trivial but quite common problem, so do check this before you start to record a spectrum.

You can't solve all the other problems quite so easily. In older instruments the inaccuracies in printing chart paper, the deterioration of the instruments chart drive, slackness of pulleys and general wear and tear, require a different strategy. This involves calibration of the spectrum using a standard. Polystyrene film is now universally used because of its convenience to handling and because it has a large number of sharp bands. Two methods of calibration are as follows.

A spectrum is calibrated before removing the chart from the instrument, by substituting the polystyrene standard for the sample and superimposing selected bands of known frequency on the sample spectrum.

For greater accuracy calibration against a range of standards is recommended, see IUPAC publication, *Tables of Wavenumbers for the Calibration of Infrared Spectrometers*, Butterworth, 1961.

3.4.   CELLS

Alkali halide discs, liquid films and mulls are mounted between steel or aluminium plates with lugs to attach the completed assembly in the instrument beam.

Solution spectra and gases are recorded in sealed cells, usually obtained commercially.

3.4.1.   Solution Cells

Three different types of solution cells are available: fixed path-length sealed cells, semi-permanent and variable path-length cells. The latter two are illustrated in Fig. 3.4a and Fig. 3.4b.

Sealed cells cannot be taken apart for cleaning or for polishing the alkali halide plates. They can be obtained in a wide range of path lengths 0.025–1.00 mm. The shortest path lengths are ideal for volatile liquids since the cell is sealed with a septum and is leak-proof.

Semi-permanent cells are demountable so that windows can be cleaned. The spacer, usually made of PTFE and available in var-

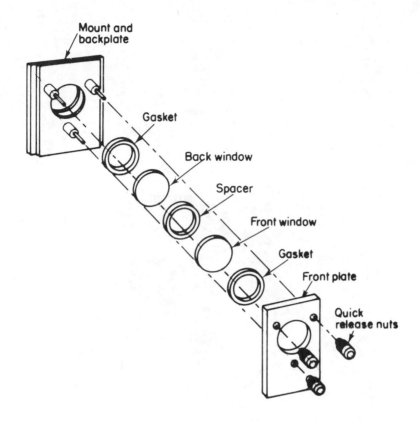

Fig. 3.4a. *A semi-permanent cell*

ious thicknesses, can also be changed, allowing one cell to be used for various path lengths: an obvious economy. These cells are not as leak-free as the bonded cells above and are not so successful for volatile liquids. Path lengths must be calibrated for accurate or quantitative work.

Variable path length cells incorporate a mechanism for continuously adjusting path length typically in the range 0–6 mm. A vernier scale allows accurate adjustment, but path-length calibration is necessary for accurate work (our old enemy water, again).

All these cell types are filled using a syringe, the syringe ports being sealed with PTFE plugs before the spectrum is recorded.

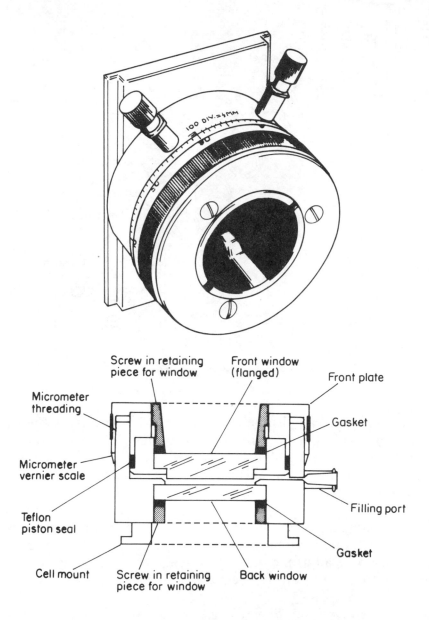

Fig. 3.4b.   *A variable path-length cell*

SAQ 3.4a
> Which type of solution cell (ie variable path-length, demountable or permanent) would you consider to be the easiest to maintain?

3.4.2. Path Length Calibration

Cell path length is measured by the method of counting interference fringes. If an empty cell with parallel windows is placed in the spectrometer and a wavelength range scanned, an interference pattern similar to Fig. 3.4c will be obtained.

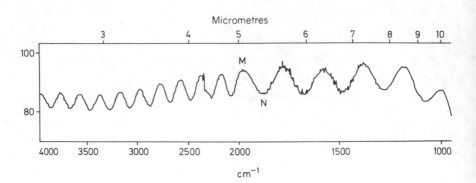

Fig. 3.4c.   *An interference pattern recorded by scanning from 4000 to 1000 cm*$^{-1}$ *with an empty cell in the sample beam*

The amplitude of the waveform will vary from 2% to 15% depending on the state of the windows. The relationship between the path-length of the cell, L, and the peak to peak fringes is given by

$$L = \frac{n}{2(\bar{\nu}_1 - \bar{\nu}_2)} \text{ cm} \tag{3.1}$$

where n is the number of complete peak to peak fringes between two maximum (or minima) at $\bar{\nu}_1$ and $\bar{\nu}_2$.

If the spectrometer is calibrated in wavelength Eq. 3.1 may be converted to a more convenient form.

$$L = \frac{n(\lambda_1 \times \lambda_2)}{2(\lambda_1 - \lambda_2)} \text{ cm} \tag{3.2}$$

where the values of wavelength, λ, are expressed in cm.

In order to understand how these fringes arise let us look at what happens to a beam of radiation as it passes through an empty cell.

When a beam of radiation is directed onto the face of a cell most will pass straight through (beam A). Some of the radiation will undergo a double reflection (beam B) and will thereby have travelled an extra

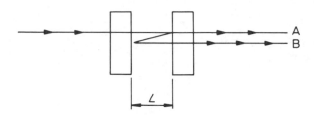

Fig. 3.4d. *A beam of radiation passing through an empty cell*

distance 2L compared with beam A. If this extra distance is equal to a whole number of wavelengths then beams A and B will be in phase and the intensity of the transmitted beam (A + B) will be at a maximum. The intensity will be at a minimum when the two component beams are half a wavelength out of phase. I think you will now be able to understand the origin of the fringes and possibly be in a position to derive Eq. 3.1 should you so wish.

SAQ 3.4b   Using the interference pattern below, calculate the path-length of the cell.

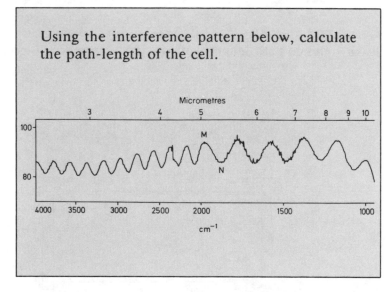

SAQ 3.4b

3.4.3.   Gas Cells

Gases have densities two to three orders of magnitude less than
liquids, hence path-lengths must be correspondingly greater. Good

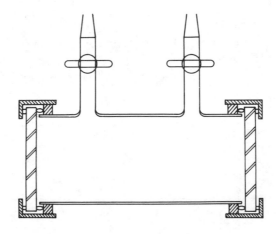

Fig. 3.4e.   *A typical gas cell*

spectra can be obtained at path-lengths of ten cm and pressures of one atmosphere or less. The volume of such a cell is of the order of 100 cm^3. The walls are of glass or brass with the usual choice of windows and the cell can be filled by flushing or from a gas line.

These cells can detect impurities down to 5%, depending on the impurity. To analyse complex mixtures and trace impurities, longer path-lengths are necessary. Typical applications would be atmospheric pollution studies, gas purity determinations and free radical or reaction intermediate studies. Commercial cells are sold with path lengths up to 20 m. This is achieved by multiple reflection optics within the cell (sample compartments are not as long as 20 m).

3.5. MORE SPECIALISED METHODS

Methods are available for materials which will not give a normal transmittance spectrum such as opaque polymers, paints, varnishes, cloth, glues, tars etc. Special methods have also been developed for micro samples from glc and hplc columns. Infrared spectroscopy can also be used to study thermodynamics and kinetics and study frozen conformations by variable temperature techniques.

3.5.1. Variable Temperature Methods

Gas cells can be obtained to allow temperatures up to 250 °C. These are used to obtain vapour phase spectra of liquids and even volatile solids.

Variable temperature cells can be obtained which are controlled to 1 °C in the range −180 to +250 °C. An electrical heating system is used for temperatures above ambient, and liquid nitrogen with a heater for low temperatures. These cells can be used for example to study phase transitions and crystallinity in polymers and kinetics of reactions.

3.5.2. Reflectance Methods

Reflectance techniques are used for samples which cannot be analysed by the normal transmittance method or which are attached to reflecting surfaces. Examples include plastic coatings, lacquers on wood, metal and paper, cloth and glues. There are three methods: Specular Reflectance, Attenuated Total Reflection (ATR) and Multiple Internal Reflection (MIR). MIR and ATR are similar, with MIR giving more intense spectra from multiple reflections, as will be seen below.

Specular Reflectance

This technique measures the radiation reflected from a surface. The material must therefore be reflective or be attached to a reflective backing. An example of the first type would be a silicon or germanium wafer, and of the second, a polymer on the surface of a christmas tree decoration.

Reflection spectra obtained from wafers, are different from, but similar to true absorption spectra, while true absorption spectra are obtained from the second type. Percentage reflection, related to a standard built into the commercial cells, is measured. Accessories are marketed to allow spectra to be obtained from chosen small areas of a surface and from samples as small as 1 mm in diameter. An obvious application is the study of surface coating damage at the molecular level.

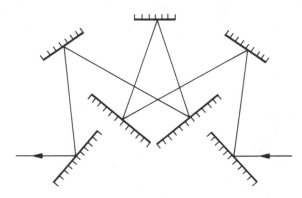

Fig. 3.5a. *Optical diagram of a micro specular reflectance unit*

ATR and MIR

Attenuated Total Reflectance has largely been overtaken by the Multiple Internal Reflectance technique, but will be described here as an introduction to the theory since MIR is simply multiple ATR. When infrared or indeed any electromagnetic radiation travels through a medium of high refractive index and meets a boundary to a material of lower refractive index, an angle of incidence exists above which total internal reflection takes place. If this surface of the prism or crystal is coated with the sample to be studied, the refractive index difference changes if the sample absorbs. Hence the intensity of radiation totally reflected changes, and a spectrum is obtained which resembles an absorption spectrum.

A prism is usually used in ATR work. MIR uses specially shaped crystals which cause many, typically 25 or more, internal reflections. This results in more intense spectra.

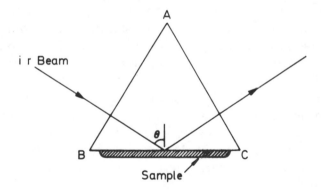

Fig. 3.5b. *Attenuated total reflectance*

The crystals used are made from materials which have low solubility in water and are of very high refractive index. Examples are KRS-5 (thallium bromide/iodide), silver chloride, germanium or vacuum deposited zinc selenide. The technique can also be used for liquids using a crystal in a cavity filled with the sample. One very significant advantage is that aqueous solutions or emulsions can be used, this saves time because it cuts out an extraction stage in the analysis.

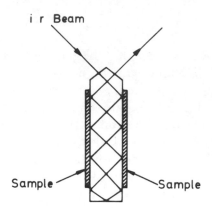

Fig. 3.5c.   *Multiple internal reflectance*

3.5.3.   Trace Samples

One of the advantages of infrared spectroscopy is the ability to handle very small amounts of sample. The standard techniques outlined above can be used for milligram quantities. Special sampling accessories are available to allow examination of microgram or microlitre amounts. This is accomplished by using a beam condenser so that as much as possible of the beam passes through the sample. A potassium bromide lens is used to focus the beam down to typically 4 × 1 mm. Microcells are available with volumes of 4 microlitres and path-lengths up to 1 mm, which are ideal for glc fraction collection. It is also possible to prepare micro alkali halide discs.

The other type of micro sample is a dilute solution which can be handled by the MIR technique above.

The fairly recent advent of cheap computer memory has allowed multiple scanning with addition of spectra in memory. Noise is random while absorptions add to give increased signal to noise ratios. The enhancement is proportional to the square root of the number of scans and hence is 100 scans gives an improvement of 10:1. This is a time consuming operation when using a dispersive spectrometer and overnight operation is usually employed. On the other hand

Fourier transform spectroscopy allows multiple scans in a short time and supersedes the conventional techniques.

3.5.4. Matrix Isolation

Matrix isolation is a low temperature technique which allows the trapping of molecules as isolated species in an inert 'matrix'. This is usually a solidified inert gas, either argon or krypton. The species to be studied is mixed with krypton or argon, the ratio of sample to matrix molecules being typically 1 : 1000, and sprayed from a gas line onto a caesium iodide window cooled to 8 K. Intermolecular interactions are therefore reduced.

The technique has had many applications including the study of reaction intermediates and radicals. Here the low temperature and concentration employed restricts possible reactions to those requiring less than a few kJ of activation energy.

The technique can also be used to study rotational isomerism around single C—C bonds in organic molecules. The low temperatures lead to very sharp bands and high resolution.

SAQ 3.5a

Why do you think low temperatures lead to sharper bands in an infrared spectrum?

(*i*) Because the concentration of the sample is very low.

(*ii*) Because there is no solute–solute interaction present.

(*iii*) Because there is no rotation at these temperatures.

SAQ 3.5a

It follows from the answer to this question that bands from trans and gauche isomers can be resolved.

By varying the temperature (8–35 K in argon) thermodynamic data can also be calculated, eg the energy barriers between rotational isomers.

Association in alcohols can be studied by 'softening' the matrix, ie allowing the temperature to increase so that the molecules can travel through the matrix. The study of hydrogen bonding is an obvious application.

The technique has also been used to assign bands in simple molecules. The spectra obtained at these low temperatures have fewer bands and are therefore easier to assign.

The technique has all the above advantages but suffers many disadvantages, the major one being cost and the difficulty of sample handling. The occurrence of sharp bands decreases as the size of the 'guest' molecule increases, the technique can only be applied to compounds having a reasonable vapour pressure at room temperature.

Summary

This Part of the Unit first covered the usual form of presentation of spectra; a plot from a recorder of % Transmittance against wavenumber in cm^{-1}.

Next the different sampling methods used when recording spectra of solids, liquids and gases were presented. The advantages of each method and problems commonly encountered in their use was then discussed.

The commonly used cells for solids, liquids and gases were briefly described and comments on their calibration and care presented.

Finally more specialised methods were presented. These included reflectance methods, matrix isolation and other variable temperature methods.

Objectives

On completion of Part Three you should be able to:

- interconvert frequency and wavelength units in the infrared region of the e.m. spectrum;

- recognise different presentations of the same infrared spectrum;

- list the advantages of infrared spectral presentation using a change of scale at 2000 cm^{-1};

- list the different methods of sample preparation and sample handling techniques for solids, liquids and gases, which are used in infrared spectroscopy;

- select appropriate sample preparation methods for different types of samples;

- recognise poor quality spectra and diagnose their causes;

- draw and recognise different types of sample cell;

- calculate the path-length of liquid cells from interference patterns;

- list uses of matrix isolation and other variable temperature methods.

4. Introduction to Spectrum Interpretation

This Part of the Unit introduces background theory which underpins the following final three Parts and will be particularly valuable in helping you with Part Seven.

We are going to start by revising and extending some important ideas that were introduced in Section 1.5 and then go on to examine factors which can be used to help you interpret infrared spectra of polyatomic molecules.

You will be aware that polyatomic molecules could have very many absorption bands in their spectra and we shall, therefore, study factors which simplify spectra. It may have struck you that detailed spectral interpretation could be a daunting task. Well you will be glad to learn that things are not as bad as they seem. We shall also study factors which complicate spectra; these can sometimes be confusing and it is important that you are aware of them before starting Part Seven.

4.1. REVISION

You should be aware that a polyatomic molecule can be looked upon as a system of masses joined by bonds with spring-like properties (see Section 1.5).

∏   Can you remember how many degrees of freedom a diatomic
       molecule has?

Well, it is logically reasonable to assume it would have twice as many
as one atom – that is six, and this is a correct assumption.

Diatomic molecules have three degrees of *translational* freedom so
what about the other three degrees of freedom. How are they used?
Well, I hope you realise that molecules also rotate and vibrate.

For diatomic molecules, rotation is meaningful about two axes only.
Hence for H—Cl there is no detectable energy involved in rotation
about the axes along the bond connecting the hydrogen and chlorine
atoms. There are therefore two degrees of *rotational* freedom. That
leaves one to be used up.

The atoms in molecules can also move relative to one another; that is
bond lengths can vary or one atom can move out of its present plane.
This is a description of stretching and bending movements which are
collectively referred to as vibrations. For our simple example, only
one vibration is possible which corresponds to the stretching and
compression of the H—Cl bond. There is therefore one degree of
vibrational freedom and all six are now used up.

Polyatomic molecules containing many (N) atoms will have $3N$ degrees of freedom. You have already seen that we can distinguish two
groups of triatomic molecules, linear and non-linear, two simple examples being CO_2 and H_2O

O
/ \
H   H
Non-linear   Linear

O=C=O

∏   Can you remember how many degrees of vibrational freedom
       water and carbon dioxide have?

If you found that at all difficult then have a look again at Section
1.5.

All polyatomic molecules have $3N$ degrees of freedom, where N is the number of atoms in the molecules. Each of the simple molecules above has three degrees of translational freedom. Water has three degrees of rotational freedom but the linear molecule carbon dioxide has only two, since no detectable energy is involved in rotation around the $O=C=O$ axis. Subtracting these from $3N$ we have $3N - 5$ for carbon dioxide (or any linear molecule) and $3N - 6$ for water (or any non-linear molecule). N in both examples is three, so, CO_2 has four vibrational modes and water has three.

Let us apply this to some more complex molecules.

SAQ 4.1a

How many vibrational degrees of freedom do the following molecules possess?

(i) Methane (CH_4)

(ii) Ethyne ($CH{\equiv}CH$)

Two other concepts introduced in Section 1.2 were used to explain the frequency of vibrational modes. You should recall that these were the stiffness of the bond, represented by the force constant, and the masses of the atoms at each end of the bond.

SAQ 4.1b

(*i*) Would you expect the bonds C—N, C=N and C≡N to vibrate at different frequencies when stretched and released?

If so which would vibrate with the (*a*) highest and (*b*) lowest frequency?

(*ii*) Which bond would vibrate (stretch and compress) at the highest frequency – a carbon–hydrogen or carbon–chlorine single bond?

The equation relating force constant, reduced mass and absorption frequency is given below;

$$\nu_e = \frac{1}{2\pi}\sqrt{\frac{f}{\mu}} \tag{1.7}$$

where ν_e is the frequency of vibration, f is the force constant of the bond and μ is the reduced mass.

You will recall that the reduced mass may be expressed as

$$\frac{1}{\mu} = \frac{1}{m_1} + \frac{1}{m_2} \tag{1.8}$$

where m_1 and m_2 are the masses of the atoms at the ends of the bond.

In a few minutes you may find it useful to use the following alernative equation.

$$\mu = \frac{m_1 m_2}{m_1 + m_2} \tag{4.1}$$

A molecule can only absorb radiation when the incoming infrared radiation is of the same frequency as one of the fundamental modes of vibration of the molecule.

This means that the vibrational motion of a small part of the molecule is increased while the rest of the molecule is left unaffected.

Before we attempt a short calculation let us modify Eq. 1.7 so that we can make direct use of the wavenumber values for bond vibrational frequencies.

$$\bar{\nu} = \frac{1}{2\pi c}\sqrt{\frac{f}{\mu}} \tag{4.2}$$

where c is the speed of light.

SAQ 4.1c   Given that the C—H stretch vibration for chloroform occurs at 3000 cm^{-1}, calculate the C—D stretch frequency for deuterochloroform.

That was fairly easy, let's try something a little more complex.

SAQ 4.1d   Assuming that the force constants for C≡C, C=C and C—C are in the ratio $3:2:1$, and that the normal range for the C=C stretch absorption is 1630–1690 cm^{-1}, what range would you expect for the C—C stretch and C≡C stretch absorptions?

SAQ 4.1d

Even if you struggled with that question I think you will agree that
we have a classical equation that describes stretching vibrations reasonably well.

Vibrations can of course involve either a change in bond *length*
(stretching) or bond *angle* (bending).

SAQ 4.1e   Draw diagrams to represent:

(*i*)   the stretching vibration of the O—H bond in an alcohol, ROH

(*ii*)   the bending vibration of the carbon skeleton of propane.

(*iii*) the stretching vibration of the C≡N bond in the nitrile RCN.

Let us consider the water molecule again. In addition to the stretching vibration already introduced above, the two O—H bonds can stretch in-phase, ie

This is *symmetrical* stretching while that for the alcohol in the previous question was *antisymmetrical* stretching. The form of stretching vibrations are easy to visualise and therefore fairly straightforward. Let's now look at bending vibrations in a little more detail. We shall use the molecule bromochloromethane (H_2CBrCl) as the example. It is best to consider the molecule being cut by a plane through the hydrogen atoms and the carbon atom. The hydrogens can move in the same direction or in opposite directions in this plane, here the plane of the paper.

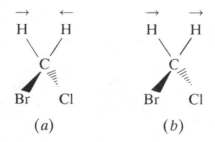

Case (a) is usually referred to as a deformation and (b) as a rock. There are two more bending vibrations out of this plane. They are referred to as (c) wags and (d) twists.

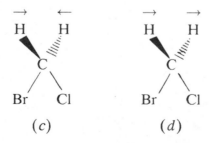

To summarise:

(*a*) is a deformation,
(*b*) is a rock,
(*c*) is a wag,
(*d*) is a twist.

It is usual to use a simpler notation than the one above. Use +
to indicate movement *up* from the plane of the paper and − for
movement *down*, (compare Fig. 1.5b).

For the four vibrations above we have:

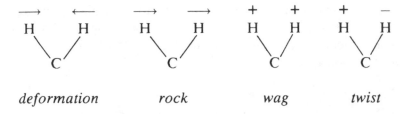

deformation   rock   wag   twist

SAQ 4.1f

(*i*) Consider the CH_2 fragment of the molecule
propane, $CH_3CH_2CH_3$. How many different bending vibrations can occur – 4, 5 or
6?

(*ii*) Now consider water. What is the total number of bending vibrations? Using the →, +
and − notation, write them down.

SAQ 4.1f

For more complex molecules the analysis often becomes simpler since we can often consider hydrogen atoms in isolation because they are usually attached to more massive and therefore more rigid parts of the molecule.

out-of-plane bend   in-plane bend

However there will be very many different vibrations for even fairly simple organic molecules. Let's look at the allyl bromide molecule:

$$CH_2{=}CHCH_2Br$$

Applying our formula to this non-linear molecule we could analyse $3N - 6$, that is 21 different vibrations.

These are:

- — C=C stretch, C—H stretch, C—C stretch, C—Br stretch,

- — in-plane and out-of-plane bending of the terminal CH_2 group,

- — in plane and out-of-plane bending of the =CH group,

- — deformation vibrations of the CH_2Br group,

- — rocks and wags of the same group,

- — twisting vibrations about the C=C and C—C bonds,

- — bends of the C—C—C skeleton.

Hence, as you are beginning to appreciate, even for a relatively simple molecule there are lots of absorption peaks in the infrared spectrum!

4.2. FACTORS WHICH SIMPLIFY INFRARED SPECTRA

In practice the complexity of practical ir spectroscopy is simplified by three factors.

- — The frequencies of many vibrations lie outside the normal scanning range of instruments.

- — By no means all vibrations give rise to strong absorption and some vibrations are ir *inactive*.

- — Some bands have very similar frequencies and coalesce. This is true for hydrocarbons which contain many similar bonds.

4.2.1. The Normal Spectral Range

You saw in Part Three that the normal range of many spectrometers is 4000–650 cm^{-1}. This will always be true for instruments with sodium chloride optics. The use of diffraction gratings has extended this range to 400 cm^{-1} or 200 cm^{-1}. This is usually available on modern instruments, but sodium chloride plates and cells cannot be used if use is to be made of the region below 650 cm^{-1}. Most laboratories use sodium chloride accessories as standard, so the 'normal' range is still looked on as 4000–650 cm^{-1}.

Remembering our classical equation above relating the frequency of absorption to force constant and reduced mass, ie

$$\bar{\nu} = \frac{1}{2\pi c}\sqrt{\frac{f}{\mu}} \tag{4.2}$$

try the following question,

SAQ 4.2a	Are the following statements true or false?

(*i*) Given that force constants for bending vibrations are considerably smaller than those for stretching vibrations and that C—O str is typically at 1200 cm^{-1}, C—O bending vibrations are likely to be outside the normal range.

(*ii*) C—F str occurs around 1400 cm^{-1}, so C—I str is likely to occur at very low frequency.

(*iii*) Vibrations of bonds involving atoms of high atomic number, eg chromium are likely to be out of range of most routine instruments.

SAQ 4.2a

∏   You have seen from this question that many low energy vi-
        brations lead to absorption below 650 cm^{-1}. Do any absorb
        above 4000 cm^{-1}?

C—H, O—H and N—H stretching vibrations absorb at or above
3000 cm^{-1}. For a bond to absorb above 4000 cm^{-1}, the reduced
mass would have to be less than C—H and the force constant greater
than that of C—H str. There are no combinations of force constant
and reduced mass that satisfy these criteria, except H_2 which is IR
inactive.

4.2.2.   To Absorb or Not to Absorb

As you already know, for a vibration to give rise to absorption of
ir radiation, it must cause a change in the dipole moment of the
molecule. The larger this change, the more intense the absorption
band will be, see Section 1.5.2.

Because of the difference in electronegativity between carbon and oxygen the carbonyl group is permanently polarised:

$$\underset{C=O}{\overset{\delta^+ \;\; \delta^-}{\diagup}}$$

Stretching this bond will increase the dipole moment and hence C=O str is an intense absorption in acids, ketones, aldehydes, acid chlorides etc.

In carbon dioxide two different stretching vibrations are possible; symmetric and antisymmetric.

$$\begin{array}{cc} \overset{\delta^- \;\; \delta^+ \;\; \delta^-}{O=C=O} & \overset{\delta^- \;\; \delta^+ \;\; \delta^-}{O=C=O} \\ \leftarrow \;\; \longrightarrow & \leftarrow \quad \leftarrow \\ (a) & (b) \end{array}$$

∏   Which one of these vibrations is ir inactive?

A dipole moment is a vector sum. CO_2 in the ground state therefore has no dipole moment. If the two C=O bonds are stretched symmetrically there is still no net dipole. However, in the antisymmetric stretch the two C=O bonds are now of different length and hence the molecule *has* a dipole. So the vibration in (b) above is ir active.

SAQ 4.2b   In contrast to the carbonyl group, the double bond of ethene $H_2C=CH_2$ is electrically symmetrical. Consequently there is *some change/no change* in dipole moment during a stretching vibration, and such a vibration *does/does not* absorb ir radiation and *is/is not* observed in the ir spectrum of the compound.

Write a few lines of explanation, then re-write the sentence above with one of the alternatives.

SAQ 4.2b

In practice this black and white situation does not prevail and shades of grey appear. The change in dipole moment may be very small and hence lead to very weak absorption.

For example the double bond in propene:

$$CH_3-CH=CH_2$$

is only weakly polarised as a result of the inductive effect of the methyl group, with the result that the $C=C$ str vibration is almost non-observable.

SAQ 4.2c   Would you expect the $C=C$ str absorption to be *active/inactive*, *weak/strong* in the following compounds. $\longrightarrow$

**SAQ 4.2c
(cont.)**

Place them in order of increasing intensity.

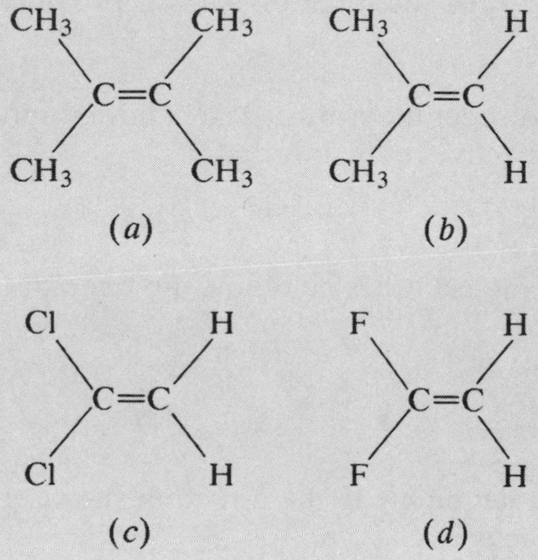

(a) (b)

(c) (d)

Bending vibrations provide more difficulties, since you may have to think in three dimensions. Let's look at some simple examples first. If we consider flat molecules and their bending vibrations in the plane, then the three dimension problem does not complicate matters.

∏   Consider the symmetrical bending vibration of CO_2. Will this be active in the infrared?

$$O=C=O$$

Here the molecule has no dipole moment, but in the bent molecule,

there is a net dipole in the direction shown. Hence the vibration is infrared active.

SAQ 4.2d   Are the following bending vibrations active or inactive?

(*i*)

(*ii*) The twisting vibration in H_2S

(*iii*) H—C≡C—H

(*iv*) H—C≡C—H

SAQ 4.2d

Π Predicting the intensity of these absorptions is even more difficult. Consider the two C—H out-of-plane bending vibrations below:

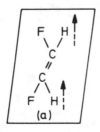 

(a)   (b)

Both are infrared active, but which would be more intense?

In (*a*) the change in dipole is in the same direction for both bonds and is additive. In (*b*) there is still a change but it is subtractive in the direction of the overall dipole. The change in dipole moment is therefore larger for (*a*), so that the intensity of the absorption from (*a*) should be greater than from (*b*).

Similar conclusions can be reached for benzene derivatives. Consider the three compounds below:

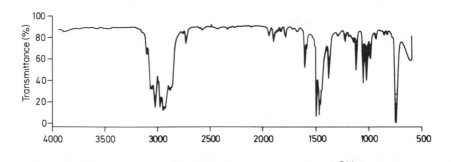

(a)   (b)   (c)

There are obviously bending and stretching vibrations of the C—H bonds in these molecules which are infrared active. They would be complex to analyse, and we shall not attempt it! The important point is that they will be different for the three different substitution patterns and for any other pattern for more or less highly substituted derivatives.

These bands lead to a series of peaks at 3000–3150 cm^{-1} for C—H str, 1200–1500 cm for in-plane bending and 650–850 cm^{-1} for out-of-plane bending vibrations. The pattern varies for different

Fig. 4.2a.   *Infrared spectrum of* C_8H_{10}:

substitution patterns and is used for structure determination (see Part Seven). Examine the spectra (Fig 4.2a, Fig 4.2b and Fig 4.2c) below and convince yourself that differences in molecular structure lead to large differences in spectra.

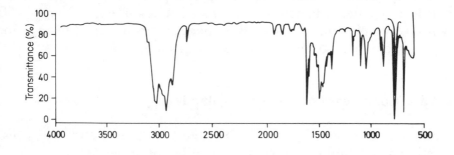

Fig. 4.2b. *Infrared spectrum of* C_8H_{10}:

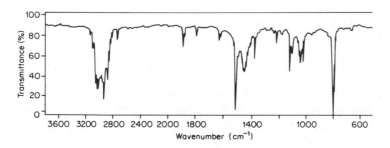

Fig. 4.2c. *Infrared spectrum of* C_8H_{10}:

These complex spectra are difficult to analyse in detail. Some general conclusions can be put forward, however, now that thousands of spectra have been studied.

Symmetrical molecules will have fewer infrared active vibrations than unsymmetrical molecules. This leads to the conclusion that symmetric vibrations will generally be weaker than antisymmetric vibrations, since the former will lead to a small change in dipole moment. It will also be true that the bending or stretching of bonds involving atoms in widely separated groups of the periodic table, will lead to intense bands. Vibrations of bonds like $C-C$ or $N=N$ will give weak bands. This again is because of the small change in dipole moment associated with their vibrations.

4.2.3.   Coalescence of Absorption Bands

You have seen that groups like CH_2 and CH_3 give rise to two types of stretching vibrations – symmetric and antisymmetric.

∏   How many peaks, in the region 2800–3100 cm^{-1} would you expect from the following compounds, the samples being liquid films?

   (*a*) $(CH_3)_3CCH_2CH_2CH_3$

   (*b*) $CH_3COCH_2COCH_3$

   (*c*) $Si(CH_2CH_3)_4$

I suppose at least four from each and maybe more from (*a*) since the CH_2's are not equivalent. With a routine instrument you would be lucky to see three. Under high resolution four would be seen.

Typical frequencies would be,

   CH_3 2962 and 2872 cm^{-1} CH_2 2926 and 2853 cm^{-1}

Remember that these bands would not be sharp, since they are envelopes enclosing many unresolved peaks.

Many bending vibrations vary little in force constant and hence cannot be resolved, especially in the liquid or solid phase.

This conclusion can be extended to most organic groups leading to fewer bands than theoretically predicted for large molecules. It is found that molecules of similar structure give similar, though not identical, spectra.

Cyclohexanone contains five CH_2 groups but only two different C—H str frequencies. This is because each CH_2 group vibrates essentially in isolation of the rest of the molecule so that all CH_2 vibrations are of similar energy.

This is the foundation of the empirical method for interpretation of infrared spectra. If theoretical calculations had to be completed for each spectrum recorded, then ir would not be as popular as it is!

SAQ 4.2e	Would you expect the spectrum of polystyrene to be similar to the spectrum of:

(i) isopropylbenzene   : $PhCH(CH_3)_2$

(ii) styrene   : $PhCH{=}CH_2$

(iii) 1,3-diphenylethane : $PhCH_2CH_2CH_2Ph$

SAQ 4.2e

To summarise, we have three simplifying factors, as follows:

— The frequencies of many vibrations lie outside the normal scanning range of instruments.

— By no means all vibrations give rise to strong absorption, some vibrations are ir *inactive*.

— Some bands have very similar frequencies and coalesce. This is true for hydrocarbons which contain many similar bonds.

To conclude this Section try the following question.

SAQ 4.2f   Consider the vibrational properties of *trans*-2,3-diiodo-2-butene

$$CH_3 \qquad\qquad I$$
$$\diagdown C{=}C \diagup$$
$$I \diagup \qquad\qquad \diagdown CH_3$$

Illustrate each of the three factors that can lead to simplification of infrared spectra.

SAQ 4.2f

4.3. FACTORS WHICH COMPLICATE SPECTRA

So far in the text we have been looking at infrared spectra in terms of the bonds giving rise to particular vibrations. It is however not true that all bands in a spectrum can be associated with a particular bond or part of the molecule. Other complicating factors have to be taken into account.

Here I would like to discuss these factors and how some of them can be put to good use.

These are:

(*a*) Overtone and combination bands,

(*b*) Fermi resonance,

(*c*) Hydrogen bonding and other intermolecular interactions,

(*d*) Transitions difficult to assign.

4.3.1. Overtone and Combination Bands

You may be aware that no musical instrument produces a note that is of one pure frequency. Instead the sound we hear is a mixture of harmonics. This is the fundamental frequency mixed with multiples of this frequency. Overtone bands in the infrared are analogous ie they are multiples of this fundamental absorption frequency.

You should be able to answer the following question in terms of the theory you have looked at in Part One of this Unit.

∏   2-Hydroxybenzaldehyde has a strong C—H bending absorption at 1390 cm⁻¹. A weak absorption is also present at 2780 cm⁻¹. This is not a C—H str absorption, can you explain this?

This can be explained by a transition of twice the energy of the fundamental absorption transition.

Fundamental   lst overtone   2nd overtone

The energy required for the lst overtone is therefore twice the fundamental, assuming evenly spaced energy levels.

Since energy is proportional to the frequency absorbed and this is proportional to the wavenumber scale, the first overtone will appear in the spectrum at twice the wavenumber.

Combination bands arise when two fundamental bands absorbing at $\bar{\nu}_1$ and $\bar{\nu}_2$ absorb energy simultaneously. The resulting band will appear at $(\bar{\nu}_1 + \bar{\nu}_2)$ wavenumbers.

SAQ 4.3a	A molecule has strong fundamental bands at the following frequencies:

C—H bend at 730 cm^{-1}

C—C str at 1400 cm^{-1}

C—H str at 2950 cm^{-1}

Write down the frequencies (in wavenumbers) of the possible combination bands and the first overtones.

As you should know from the earlier parts of the Unit, the energy levels are not exactly equally spaced, with the difference in energy decreasing as more excited levels are reached. This approximation is, however, good enough to enable you to find the part of the spectrum to look at.

∏   Would you expect overtone bands to absorb at higher or lower wavenumber than expected, using our approximation?

If the second energy difference is slightly less than the first, then the total energy for the first overtone band will be less than our approximation, this means that the band will appear at *lower* wavenumber.

These bands are forbidden according to simple quantum mechanical analysis of the system. This means that they will be weak in comparison with fundamentals. It is therefore usually easy to pick out fundamental bands, especially for simple molecules. But beware, C—H bending and C=C str vibrations of weakly polar molecules are usually very weak because of the small dipole moment change.

SAQ 4.3b

In the table below I list the bands in the spectrum of SO_2. From the intensities and frequencies can you spot the fundamentals, overtones and combination bands?

Frequency (cm^{-1})	Intensity
519	very strong
1151	very strong
1361	strong
1871	very weak
2305	very weak
2499	moderate

SAQ 4.3b

4.3.2. Fermi Resonance

This is a very confusing effect for the unwary! The Fermi resonance effect usually leads to two bands close together when only one is expected.

I now refer back to Section 4.3.1 above, and in particular to the first question on overtones. I suggested that there was a weak overtone band in the spectrum of 2-hydroxybenzaldehyde (salicylaldehyde) at 2780 cm^{-1}. This is not true. There are in fact *two* bands at 2732 and 2833 cm^{-1}, approximately 50 cm^{-1} either side of the value I gave. The C—H str of the aldehyde HC=O group is expected at this frequency (2780 cm^{-1}) in addition to the overtones.

When an overtone or combination band has the same or similar frequency as a fundamental, two bands appear split either side of the expected value and of reasonably equal intensity. The effect is greatest when the frequencies match, but is present when there is a mismatch of a few tens of wavenumbers. The two bands are referred to as a 'Fermi doublet'.

SAQ 4.3c	Tetrachloromethane is expected to show only four infrared active fundamentals. Three fundamentals absorb at 217 (infrared), 313 (infrared and Raman) and 459 cm^{-1} (Raman only). The fourth is expected to occur in the region 700–800 cm^{-1}. The spectrum has two bands in this frequency range, at 762 and 791 cm^{-1}. Can you account for this observation?

Unfortunately things are even more complex. For Fermi resonance to occur two further conditions have to be met:

— the fundamental and overtone should arise from the same part of the molecule,

— they must have the same symmetry properties.

The second point is difficult to explain without a lot more theory, but if you keep to in-plane symmetrical or antisymmetrical vibrations you will not go far wrong.

Hence in the salicylaldehyde example above the two bands must have the same symmetry hence the overtone must be an in-plane bending vibration, since the fundamental is the C—H str of the HC=O group.

SAQ 4.3d

Label the following statements as referring to overtone, combination or Fermi resonance bands.

(i) 'Carbonyl stretching bands sometimes are doublets separated by as much as 50 cm^{-1}.'

(ii) 'The series of bands in benzene derivatives between 1800 and 2000 cm^{-1} are dependent on the substitution pattern in the ring.'

(iii) 'The precise position of this band depends on the frequency of two other bands at lower frequency.'

4.3.3. Hydrogen Bonding and Intermolecular Interactions

Spectra of the same compound may be different when examined in different solvents or in the solid phase rather than in solution. Both will be very different from a vapour phase spectrum. Most organic molecules are polar. There are therefore interactions between dipoles in the solid state which can shift some absorption bands and broaden others. Similar interactions will be present in a solution of the compound in a polar solvent. Hydrogen bonding will be dealt with in detail in Part Five, and this again leads to major changes in spectra.

With the introduction of microcomputers it is now possible to compare a spectrum against a library and search for best matches as an aid in compound identification. This search can also be done manually using published catalogues.

The above effects should be taken into account in these searches, aways note the conditions under which the spectra were obtained.

The only sure way is to compare spectra from the same instrument under the same sampling conditions, which is of course impossible.

Compare the two spectra of the same compound run under different conditions, as shown in Fig. 4.3a (a mull) and Fig. 4.3b (in solution). There are large differences.

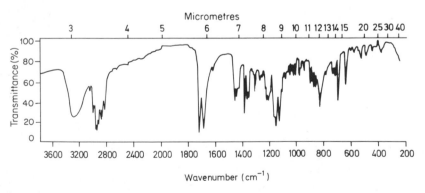

Fig. 4.3a.   *Mull*

Note the splitting of the carbonyl peak at 1700 cm^{-1}, the region above 3000 cm^{-1}, etc.

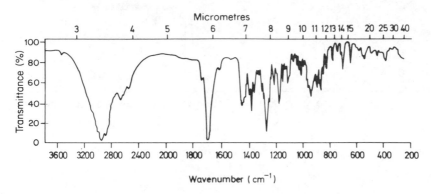

Fig. 4.3b. *Solution*

4.3.4. Transitions Difficult to Assign

We have been assuming in the discussion above that each band in an infrared spectrum can be assigned to a particular deformation of the molecule, the movement of a group of atoms or the bending or stretching of a particular bond. This is possible for many bands, particularly stretching vibrations of multiple bonds which are 'well-behaved'. However many vibrations are not well behaved and may vary by hundreds of wavenumbers in similar molecules. This applies to most bending and skeletal vibrations, which absorb in the 1500–650 cm^{-1} region, for which small steric or electronic effects in the molecule lead to large shifts. A spectrum of an organic compound may have a hundred or more absorption bands present, but there is no need to assign the vast majority of them. The spectrum can be looked on as a 'fingerprint' of the molecule.

Look at the region below 1500 cm^{-1} in the spectra of 2-methyl and 3-methylpentane given Fig. 4.3c and Fig. 4.3d.

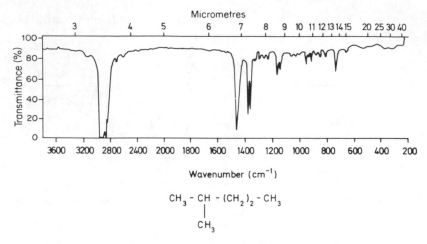

Fig. 4.3c. *Infrared spectrum of 2-methylpentane*

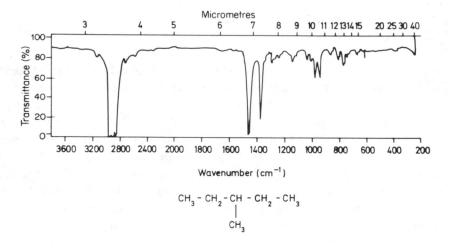

Fig. 4.3d. *Infrared spectrum of 3-methylpentane*

Let's now look at this region in more detail, pointing out bands that can be assigned.

4.4. THE FINGERPRINT REGION

You should also recall that absorption frequency is proportional to
the square root of the force constant of the bond, which is related
to bond strength.

∏   Assuming that the force constants are roughly equal, would
     you expect C—O, C—N and C—C bonds to have very dif-
     ferent stretching frequencies?

Let's work out the reduced masses. Using μ_{CO}, μ_{CN} and μ_{CC} for the
three reduced masses, (omitting the Avogadro number).

$$\mu_{CO} = \frac{12 \times 16}{12 + 16} = 6.9; \qquad \mu_{CN} = \frac{12 \times 14}{12 + 14} = 6.5;$$

$$\mu_{CC} = \frac{12 \times 12}{12 + 12} = 6.0$$

The square roots of the reciprocals of these numbers are:

C—O 0.38;   C—N 0.39;   C—C 0.41

Therefore assuming our force constant approximation is a good one, there is not going to be much difference in vibrational frequency, say 10%.

They are found to absorb in the following ranges:

C—O 1000–1300 cm^{-1};   C—N 1000–1380 cm^{-1};

C—C 750–1200 cm^{-1}

As you can see, they are mutually overlapping, and also cover large ranges. They vary wildly for small changes in molecular shape and are therefore not used for structure determination, as we shall see in Part Seven.

Bending vibrations involving these atoms are likely to be outside the range of routine instruments. If you recall the difference in frequency between C—H stretching and bending vibrations and impose a similar ratio here you will see what I mean.

Typical C—H stretch,   3000 cm^{-1}

Typical C—H bend,   1000 cm^{-1}

C—O, C—N and C—C bending are therefore likely to absorb at less than 400 cm^{-1}. These bending vibrations again occur over a wide range and vary significantly for small changes in structure. These modes (both bending and stretching) are referred to as skeletal vibrations.

There is another reason why absorption bands below about 1500 cm^{-1} are difficult to assign.

Vibrations in the skeleton become coupled. In other words they are not restricted to one or two bonds, but may involve a large part of the carbon backbone and oxygen or nitrogen atoms if present. The energy levels mix, resulting in the same number of vibrational modes, but at different frequencies. Bands can no longer be assigned to one bond. This is very common and occurs when adjacent bonds have closely similar frequencies. Coupling commonly occurs between C—C, C—O, C—N stretch and C—H rocking and wagging motions. A further requirement is that to be strongly coupled the motions must be in the same place.

These are restrictions which you have met above. Can you remember the context? You met similar laws for Fermi resonance to be allowed.

There are then three factors which determine whether coupling will take place:

— The groups must be adjacent.

— The frequencies must be similar.

— The vibrations must be in the same plane.

SAQ 4.4b   Select from the list below the vibration of butane ($CH_3CH_2CH_2CH_3$) which you would expect to couple strongly with the C—C stretching modes. Remember that there are three factors which determine whether vibrational coupling will take place.

(*i*)   C—H stretch

(*ii*) CH_3 twist

(*iii*) CH_2 rock

(*iv*) CH_2 wag

(*v*)   none of these

SAQ 4.4b

It turns out that all vibrations which absorb in the region 1500–650 cm^{-1} will couple to an extent that critically depends on the shape of the molecule. Large groups, the presence of double and triple bonds and even electronic effects will influence this. Hence it is very difficult to predict the extent of this coupling. This leads to great difficulties in the assignment of bands in this region of the spectrum. This then accounts for the wide range over which C—C, C—O and C—N stretching vibrations absorb. On the other hand we can identify vibrations that do not couple (the 'well-behaved' ones above). We shall be using these frequencies in Part Seven of this Unit.

The most important consequence of coupling, however, is that small changes in structure leads to major changes in the infrared spectrum in the 1500–650 cm^{-1} region. It is *very* unlikely therefore that two different molecules will have exactly the same pattern of bands (run under identical conditions) in this region. This region is therefore referred to as the 'fingerprint region'.

As a parting thought, however, compare the spectra of the two straight chain hydrocarbons shown in Fig. 4.4a and Fig. 4.4b.

As molecules get larger small differences in structure have little effect on the spectrum and we would expect the spectrum of $C_{16}H_{34}$ to be very similar.

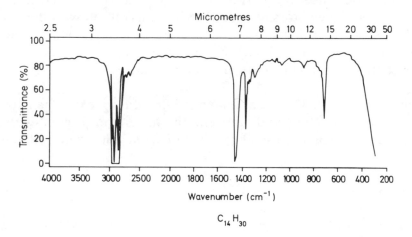

Fig. 4.4a. *Infrared spectrum of the straight chain hydrocarbon,* $C_{14}H_{30}$

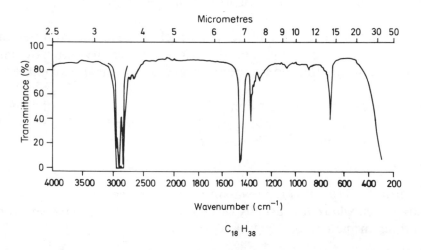

Fig. 4.4b. *Infrared spectrum of the straight chain hydrocarbon,* $C_{18}H_{38}$

4.5.   RELIABLE ABSORPTIONS: A SUMMARY

You should appreciate from the above text that many ir absorptions are highly unreliable in that they can be of very variable frequency. We have looked at the effects that lead to this sad state of affairs. However, many absorptions are highly reliable and can be used for compound identification. Part Seven of this Unit is devoted to this topic. It is sufficient to say here that bearing the above effects in mind it is possible to use ir spectroscopy as a very valuable aid, sometimes in conjunction with other techniques, in the identification of both organic and inorganic compounds.

In general, peaks that occur at wavenumber values greater than 1600 cm^{-1} are very reliable, while those in the fingerprint region, with a few exceptions, which we shall discuss in Part Seven are not reliable.

Summary

This Part of the Unit started with the classification of molecular vibrations for simple diatomic and triatomic molecules and then moved on to discuss complex polyatomic molecules. The latter were shown to exhibit very many possible modes of vibration which give rise to a large number of absorption bands in their spectra.

Several factors were then discussed which reduce the number of observed bands. These included the infrared inactivity of particular vibrations and the limited frequency range of typical spectrometers.

A number of factors which complicate the appearance and hence the interpretation of spectra were then presented. They included overtone and combination bands, Fermi resonance and solvent effects.

The term *fingerprint region* was introduced to cover the region of the spectrum where it is usually difficult to assign ir bands to specific vibrations in molecules.

Objectives

On completion of Part Four you should be able to:

● discuss the influence of force constants and reduced masses on the frequency of band vibrations;

● state the selection rule for absorption of infrared radiation;

● draw diagrams representing bending and stretching modes of vibration;

● list the factors which simplify infrared spectra;

● understand the factors which complicate infrared spectra –

  ○ skeletal vibrations
  ○ combination and overtone bands
  ○ Fermi resonance
  ○ intermolecular interactions

● state the frequency ranges of the 'fingerprint region' and the 'reliable region' of the infrared spectrum.

5.   Hydrogen Bonding

5.1.   THE IMPORTANCE OF THE HYDROGEN BOND

Why devote a whole Part of this Unit to hydrogen bonding?

If you have studied organic chemistry in any depth you will have encountered this effect in the chemistry of alcohols, phenols, acids and perhaps amines, other nitrogen compounds and polymers.

You may have considered hydrogen bonding interesting, but not very important.

The effect is of paramount importance to us in many ways and is therefore worthy of detailed study.

One of the many techniques used to study hydrogen bonding has been infrared spectroscopy, so you can treat this Part as an example of the power of the method. The effect's importance can be exemplified by the fact that we feed and clothe ourselves with hydrogen-bonded materials, that the biological activity of DNA relies on hydrogen bonding and it also accounts for the high boiling point and many of the properties of water, without which you would not be reading this. So maybe it has some importance after all.

Within the confines of a Unit on infrared spectroscopy, the reasons for looking at hydrogen bonding in this detail are, (*a*) it is a good example of the power of infrared spectroscopy,

(*b*) it illustrates the effect of phase, solvent and structure on vibrational modes,

(*c*) it should teach you some of the dangers of interpreting spectra as the vibrations of isolated molecules,

(*d*) it should prepare you for some of the pitfalls to come in Part Six (quantitative infrared spectroscopy).

So far we have used the phrase absorption *band* rather than absorption *peak*. From now on we will not be using one or the other exclusively. The choice is largely a matter of individual preference, although strictly speaking the infrared spectra of liquids and solids exhibit bands. This is because the absorptions are made up of a series of individual peaks.

Now let's refresh your memory on hydrogen bonding.

5.2. HYDROGEN BONDING – THE BACKGROUND THEORY

Examine the data in the following table:

	Melting Point/K	Boiling Point/K
NH_3	195	240
CH_4	89	112
H_2O	273	373
H_2S	190	231
HF	181	292
HCl	161	189

How can we account for the high melting and boiling point of ammonia compared to methane, when the molecular masses of the molecules differ by only one unit?

Firstly, methane is more symmetrical than ammonia. This would tend to raise its melting and boiling points since the molecules can pack more easily in the solid state. The N—H bond is slightly more polar than the C—H bond so dipole–dipole attraction in the solid and liquid could perhaps explain some of the effect. We have to invoke another force of attraction, however, since the H—F bond is more polar than both, yet shows similar properties.

Ammonia possesses a lone pair of electrons on the nitrogen. These electrons can be shared with a hydrogen atom on another molecule of ammonia to form a 'hydrogen bond'. This sharing is eased by the small size of the hydrogen atom allowing electrostatic interactions to become important. The resulting hydrogen bond leads to dimeric and higher aggregates in the solid and liquid state, hence raising the boiling and melting points. The same argument applies to the first of each pair of compounds in the table.

There is no universally accepted definition of the hydrogen bond. One definition is based on infrared spectroscopy, others on types of group in the molecule and yet another on physical measurements like those above.

We obviously need a working definition. Let's say that a hydrogen bond exists when there is evidence of bond formation through association and involving a hydrogen atom in the same or another molecule.

The standard symbol for this bond is a dashed line, (||||||).

SAQ 5.2a   Which of the following molecules would you expect to be hydrogen bonded?

Write down a structure for the associated aggregate.

(*i*) SiH_4

(*ii*) Acetic acid (Ethanoic acid)

(*iii*) $CH_3C{=}O$
           |
           NH_2

(*iv*) Phenol

The hydrogen bonds in the molecules above are *inter*molecular, ie they occur between two or more molecules of the compound and the resulting associated species is in equilibrium with the non-associated monomer. There is a further possibility. If we look at the structure of salicylic acid (2-hydroxybenzoic acid) it becomes clear that a hydrogen bond can form between the OH and COOH groups *within the same molecule* as shown in the structural formulae below:

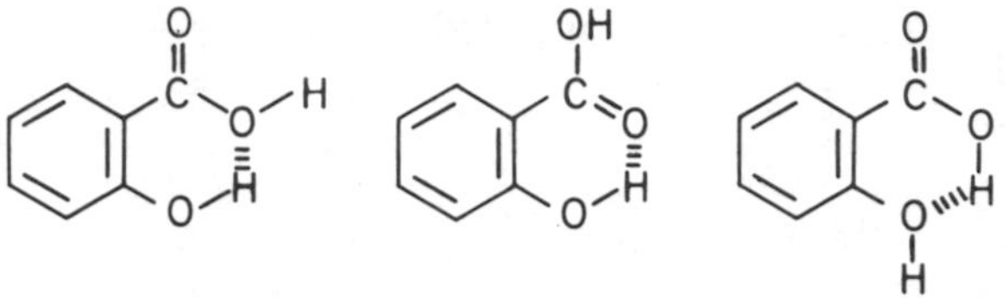

This is known as an *intra*molecular hydrogen bond. It is found that these hydrogen bonds are strongest when the resulting ring size is five, six or seven. Other ring sizes are sterically impossible or the resulting hydrogen bond weak.

∏   Would you expect intramolecular hydrogen bonding in the following molecules?

(*a*) 2-methoxyphenol

(*b*) 2-hydroxybenzaldehyde

(*c*) propane-1,3-diol

Yes, all of them. I have drawn the structures below:

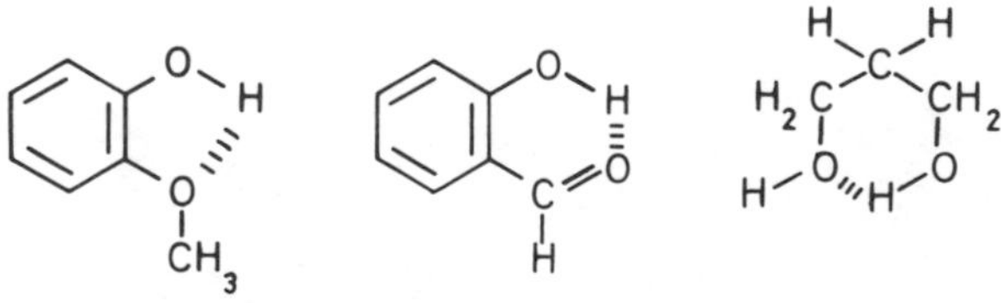

Note that a five membered ring is formed in (*a*) and six membered rings in (*b*) and (*c*).

The presence of *intra*molecular hydrogen bonds does not mean that *inter*molecular hydrogen bonds are absent, both are present.

The formation of *intra*molecular hydrogen bonds depends on the participating groups being in the correct spacial relationship to each other. By this I mean at a distance where a reasonable interaction between the hydrogen atom and the lone pair electrons on the donor atom is possible. This is also true for *inter*molecular hydrogen bonds. The groups taking part in the bond must meet.

SAQ 5.2b

Classify the following compounds as *inter*molecularly hydrogen bonded, *inter* and *intra*molecularly hydrogen bonded or lacking hydrogen bonds.

(*i*)   ethanol,

(*ii*)   2-aminobenzoic acid,

(*iii*)   chloroform,

(*iv*)   ethyl acetoacetate ($CH_3\overset{\overset{O}{\|}}{C}CH_2\overset{\overset{O}{\|}}{C}OCH_2CH_3$).

SAQ 5.2b

Example (*iii*) in the question above leads to a very important point. Many solvents are capable of forming hydrogen bonds to solutes. To form a hydrogen bond we need a proton donor group and an electron donor group. Three classes of solvent exist which may lead to trouble when used for hydrogen bonding studies in solution.

1. Compounds which contain hydrogen donor groups only, eg halogenated compounds which contain a sufficient number of halogens to activate the hydrogens present. In this class is chloroform, a very common solvent for infrared studies. If the compound under study contains non-bonded electron pairs then some effect on the spectrum is inevitable.

2. Compounds which contain non-bonded electron pairs. In this class are ethers, adehydes, ketones, tertiary amines etc. If the compound being studied contains a proton donor group, solvent effects on the spectrum should again be present.

3. Compounds which contain both types of group. Here we have water and alcohols. Hydrogen bonding to nearly all organic compounds is now possible.

The only usable solvents are those not possessing either of these features. I can only think of two common ones, CCl_4 and CS_2. These still contain lone pairs but being on S and Cl are less available, and any interaction will therefore be extremely weak.

So hydrogen bonding can occur when you are least expecting it, and it has to be kept in mind when choosing a solvent for any compound. This can often explain the differences in infrared spectra of compounds in different solvents.

5.3. EXPECTED CHANGES IN SPECTRA

In this Section I would like to explore what effect the presence of hydrogen bonding has on the infrared spectrum of a compound.

Infrared spectroscopy is an excellent source of information on molecular structure because the frequencies of vibration of bonds depend on the masses of the atoms in the bond and the bond stiffness. Any factor which influences the bond stiffness will also alter the frequency of vibration. We would expect hydrogen bonding to do this. Let's look at this in a little more detail.

Taking the hydrogen bond in an alcohol as an example, there are five new vibrational modes, in a hydrogen bonded dimer. I list them below with their usual range of absorption.

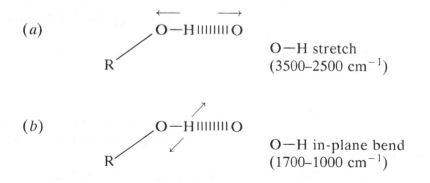

(a)   O—H ‖‖‖‖‖ O

O—H stretch
(3500–2500 cm^{-1})

(b)   O—H ‖‖‖‖‖ O

O—H in-plane bend
(1700–1000 cm^{-1})

(c) O$-$H ⅠⅠⅠⅠⅠⅠⅠ O

       R

                  O$-$H out-of-plane bend
                  (900–300 cm^{-1})

(d) O$-$H ⅠⅠⅠⅠⅠⅠⅠ O

       R

                  H......O stretch
                  (250–50 cm^{-1})

(e) O$-$H ⅠⅠⅠⅠⅠⅠⅠ O

       R

                  H......O bend
                  (<50 cm^{-1})

Modes (d) and (e) are well out of range of normal instrumentation and have been little studied, and we shall not discuss them further here. The out-of-plane bending absorption is also usually out of range. This leaves the O$-$H stretch and in-plane bend, which can be easily studied, and we will devote the remainder of this Part to them.

First let us look at band frequency and then band intensity.

∏   How would you expect the absorption frequency of the O$-$H stretching vibration to be influenced by the presence of a hydrogen bond. Would it increase or decrease?

The bond order (by this I mean a measure of whether the bond is single, double or triple or somewhere in between, ie 3 for a triple bond, 2 for a double bond and so on) in the O$-$H bond will decrease, hence less energy will be needed to stretch it, so the frequency should *decrease*.

On the other hand I would expect the energy required for the in-plane bending mode to increase, since the H is more constrained when hydrogen bonded.

Now let us turn our thoughts to band intensity and start by answering a couple of questions.

SAQ 5.3a

(*i*) Would you expect the O—H stretching mode to vary in intensity when hydrogen bonded? (You will need some of the ideas from Part Four.)

(*ii*) If so would it increase or decrease?

We have already touched on solvent effects, it should be obvious that the intensity of bands are likely to be influenced by the polarity of the surrounding solvent molecules. In hydrogen bonded molecules this is likely to be even more important.

∏   Ignoring any solvent interactions, what effect would

(*a*) concentration,

(*b*) temperature,

have on the degree of hydrogen bonding present in a particular compound?

The lower the concentration the less chance there is of two molecules of monomer colliding. Hence it would be expected that the degree of hydrogen bonding would decrease with decreasing concentration.

Increasing the temperature means that each molecule will have more energy on average and hence weak associative forces are likely to be broken. This should lead again to a lesser degree of hydrogen bonding.

If infrared spectroscopy can be used to study hydrogen bonding, temperature and concentration effects should therefore also be evident. Let's see if these predictions are seen in spectra.

5.4. OBSERVED EFFECTS OF HYDROGEN BONDING

Study the two spectra below. Fig. 5.4a is a 10% v/v solution of ethanol in CCl_4 run at a path length of 0.1 mm, while Fig. 5.4b is a 1% solution of the same compound in the same solvent at a path length of 1.0 mm. Restrict your examination to the $O-H$ stretching region, in this case, 3000–3700 cm^{-1}.

∏   Using the ideas in Section 5.3

How many peaks are there in this region?

Do their relative intensities change?

Does the total intensity or peak area change?

Assign the peaks to particular stretching modes. Explain the intensity changes.

A careful analysis of these spectra should have given you examples of most of the material in Section 5.3. Let's look at the answers to these questions in some detail.

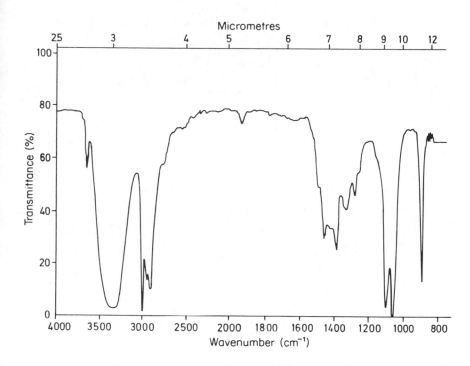

Fig. 5.4a. *Infrared spectrum of ethanol [10% v/v in CCl₄, 0.1 mm path length cell]*

The 10% solution has two peaks, one weak sharp absorption at 3640 cm^{-1}, and one strong, broad absorption centred at 3340 cm^{-1}. The 1% solution has these same two peaks. It is tempting to say that there is a third peak, the shoulder at 3500 cm^{-1}.

The sharp peak becomes more intense on dilution relative to the broad absorption. There is no sign of the shoulder in the more concentrated solution.

The total peak area decreases as the solution is diluted. Note that the peak at 880 cm^{-1} remains at the same intensity. Note also that the spectra have been run in such a way that the concentration in the beam remains constant. The solution has been diluted by a factor of ten but the path length has been increased by the same factor. The intensity of the peak at 880 cm^{-1} is a good test.

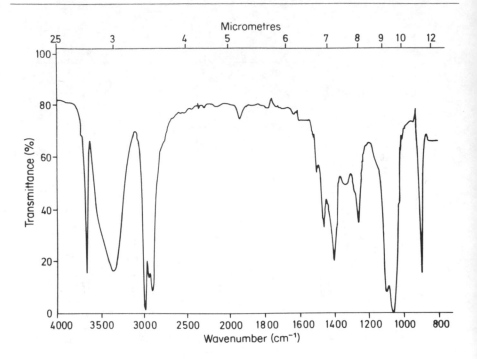

Fig. 5.4b.   *Infrared spectrum of ethanol [1% v/v in CCl₄, 1.0 mm path length cell]*

There is hydrogen bonding present. The amount of monomer or unassociated alcohol will increase as the concentration decreases. We can therefore assign the sharp band to the monomer and the broad band to polymeric material. Many workers have assigned the shoulder to dimer and the other band to higher aggregates. The sharp peak will increase in intensity at the expense of the broad peak as the concentration decreases. To explain the change in total intensity we must assume that the molar intensity of the unassociated O—H stretch is weaker than the associated stretching absorption. This is in agreement with the answer to SAQ 5.3a above.

Think about it.

You should also have noticed quite complex changes to the spectrum in the O—H in-plane bending region and the C—O stretching region.

SAQ 5.4a

Examine the ir spectra of hexan-1-ol given in Fig. 5.4c and Fig. 5.4d.

(*i*) Assign the monomer and polymer bands in the O—H stretching region.

(*ii*) How many bands are there in this region for each concentration?

(*iii*) Note any differences from the ethanol spectra you have just studied and comment on them.

The infrared spectra of carboxylic acids should show similar characteristics. Here we have the added complication of the carbonyl group, which should also be influenced by hydrogen bonding.

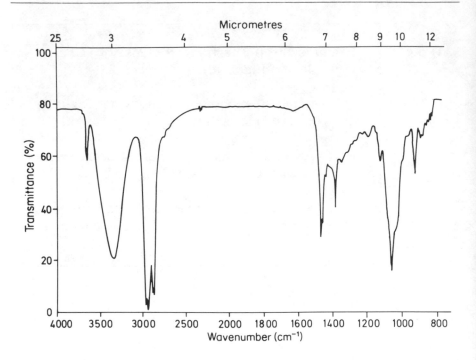

Fig. 5.4c. *Infrared Spectrum of Hexan-1-ol [10% v/v in CCl$_4$, 0.1 mm path-length cell]*

$$\overset{\longleftrightarrow}{C=O \text{IIIIIIII} H-O}$$

Hydrogen bonding here has the hydrogen between an sp^2 and sp^3 oxygen atom. We would expect the O—H stretching band to be influenced in the same manner as alcohols.

∏   What would be the effect on the carbonyl stretching frequency?

Using the same arguments we used above I would expect the bond to be weakened since the electron density on oxygen is reduced because of hydrogen bond formation. This means the force constant will be reduced and hence the absorption frequency will decrease when hydrogen bonded. In the question that follows try to use the arguments above to interpret the spectra of hexanoic acid at two different concentrations.

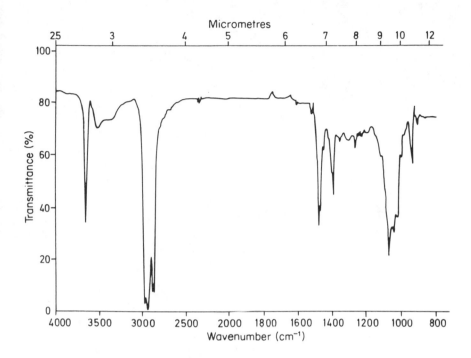

Fig. 5.4d. *Infrared Spectrum of Hexan-1-ol [1% v/v in CCl₄, 1.0 mm path-length cell]*

SAQ 5.4b

Examine the spectra of hexanoic acid given in Fig. 5.4e and Fig. 5.4f

(*i*) Assign the monomer and polymer bands in the O—H stretching region.

(*ii*) Assign the monomer and polymer bands in the C=O stretching region.

(*iii*) Compare the shape of the O—H stretching bands with those for ethanol, do you have any explanation for the differences?

SAQ 5.4b

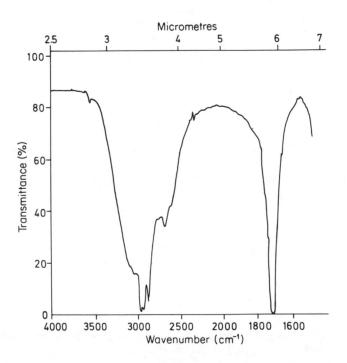

Fig. 5.4e.   *Infrared spectrum of hexanoic acid [10% v/v in CCl₄,*
0.1 mm path-length cell]

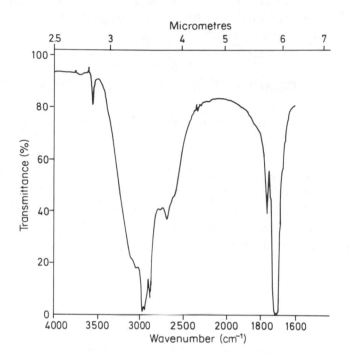

Fig 5.4f.   *Infrared spectrum of hexanoic [1.0% v/v in CCl₄, 1.0 mm path-length cell]*

∏   What effects should we see on increasing the temperature of these solutions of acids in CCl_4?

I would expect the intermolecular hydrogen bonds to break up, so that we should see an increase in intensity of the monomer peak at the expense of the polymer absorption. This is exactly what is observed.

You have seen, even at quite low concentration, that considerable aggregation takes place. It can be seen down to 0.001% concentration.

∏   Can you think of an infrared technique that could observe pure monomer spectra?

Matrix isolation in solid argon would be one technique (see Part Three). In this technique diffusion is stopped and all monomer molecules are separated by the inert gas matrix.

5.5.   INTRAMOLECULAR HYDROGEN BONDING

∏   Would you expect to be able to distinguish intramolecularly hydrogen bonded compounds using infrared spectroscopy?

If we dilute the solution enough all intermolecular hydrogen bonding will be removed; a concentration of less than 0.0005 mol dm^{-3} is usually needed to achieve this. If a broad band persists below 3600 cm^{-1} and is a constant fraction of the total O—H stretching absorption on further dilution, then intramolecular hydrogen bonding must be present.

Compare the spectra in Fig. 5.5a and Fig. 5.5b.

Both are of solutions of alcohols in carbon tetrachloride at very low concentration and show the region 4000–3000 cm^{-1} (the O—H stretching region).

It is obvious that the first exhibits little hydrogen bonding present (O—H stretch below 3600 cm^{-1}), while the other retains the hydrogen bond, even at this low concentration.

This technique can therefore be used to distinguish isomers of organic compounds and has been widely used in this way.

Typical examples are:

(*a*) The determination of the position of aromatic substitution.

        eg 2-methoxyphenol from 3- and 4-methoxyphenol.

(*b*) Distinguishing *cis* from *trans* isomers of alkenes.

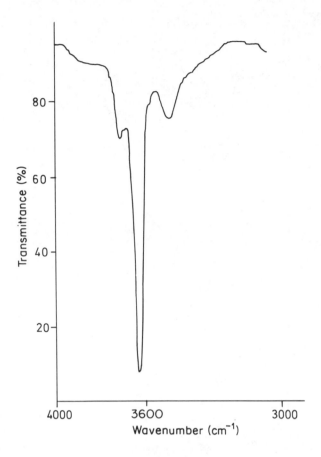

Fig. 5.5a. *Infrared spectrum of butan-1,4-diol*
(HOCH$_2$CH$_2$CH$_2$CH$_2$OH) [Very dilute solution in CCl$_4$]

eg *cis*-3-chloropropenoic acid from *trans*-3-chloropropenoic
acid.

(*c*) Conformational analysis in cyclohexane and more complex ring
systems.

1,2-diequatorial and 1,2-equatorial-axial conformers form more
stable intramolecular hydrogen bonds than 1,2-diaxial conformers, this is because the participating atoms are too far apart in
the diaxial conformers.

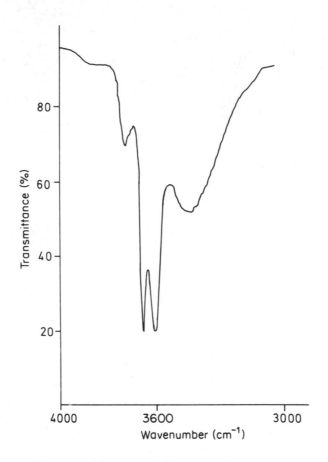

Fig. 5.5b. *Infrared spectrum of butan-2, 3-diol ($CH_3 CHCHCH_3$)*

HO OH

[Very dilute solution in CCl$_4$]

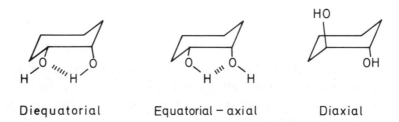

Diequatorial   Equatorial − axial   Diaxial

(*d*) The demonstration of the presence of keto-enol tautomerism in say acetylacetone (pentane-2,5-dione), which exists as an equilibrium mixture of two structures.

SAQ 5.5a

Is it possible to use the presence of an intramolecular hydrogen bond to distinguish the following pairs of isomers?

Write a structure for the intramolecularly hydrogen bonded isomer.

(*i*)   2-nitrophenol from 4-nitrophenol.

(*ii*)   trans-1,4-dihydroxycyclohexane from cis-1,4-dihydroxycyclohexane (you may need to build models of these compounds).

(*iii*) the isomers of the alkaloid, granatinol:

exo   endo

SAQ 5.5a

Attempts have been made to use use infrared spectroscopy to analyse quantitatively the mixtures of isomers present in hydrogen bonded materials. This is complex and will be studied in Part Six.

The text above should have convinced you that infrared can be used qualitatively to study hydrogen bonding. Other techniques can be used, eg comparison of melting points, deviation from the gas laws in the vapour state, nuclear magnetic resonance spectroscopy etc., but infrared is quicker (and cheaper) than most.

5.6.   EFFECT OF SOLVENT

A final word of warning. The structure of the solvent has a marked effect on the position of the delicately balanced equilibria present in solutions of hydrogen bonded materials. The detailed structure of the O—H stretching band varies for most compounds as the solvent is varied demonstrating the existence of a new set of equilibria involving the solvent. This can be demonstrated by running the spectra at very high dilution, where the effects persist. Chloroform has been shown to hydrogen bond to alcohols, behaving as the hydrogen donor to the bond. Other solvents act as the basic end of the bond. In this category we can place benzene, (through the pi-electrons), carbon disulphide, pyridine, nitromethane etc.

If pyridine is used as a solvent for hydrogen bonded materials at low concentration then the species present is mainly the hydrogen bonded complex between the compound and pyridine. This simplifies the spectra, usually giving only one band in the O—H stretching region. No bands from polymeric aggregates are seen. This technique allows the O—H stretching band to be used quantitatively (see Part Six).

So great care must be taken in qualitative and quantitative studies. Even greater care must be exercised when comparing results from different compounds. I hope I have demonstrated that concentration effects in infrared spectroscopy can often be interpreted in terms of intermolecular hydrogen bonding or interaction with the solvent. The last point should be kept in mind when choosing a solvent for infrared studies.

Summary

This Part of the Unit first placed the contents in perspective and underlined the importance of solvent effects in infrared spectroscopy.

The chemistry of the hydrogen bond was briefly revised followed by a discussion of the likely effects of hydrogen bonding on ir spectra.

Actual spectra were then studied and effects within molecules (intramolecular) differentiated from effects between molecules (intermolecular).

Finally solvent effects in infrared spectroscopy were discussed using the knowledge gained from the earlier Sections.

Objectives

On completion of Part Five you should be able to:

* explain the phenomenon of hydrogen bonding;

* write structures for dimeric intermolecularly bonded species and for intramolecularly bonded compounds;

- list and explain the expected effect of hydrogen bonding on the O—H stretching and O—H in-plane bending region of the infrared spectrum;

- give examples of the effect of temperature, concentration and solvent on the infrared spectrum of a hydrogen bonded material;

- explain how infrared spectroscopy can be used to identify an intermolecularly hydrogen bonded material;

- demonstrate an awareness of solvent choice as important in infrared studies as a result of possible solute–solvent interactions resulting from hydrogen bond formation.

6. Quantitative Analysis

Quantitative infrared spectroscopy suffers many disadvantages when compared with other analytical techniques and its use tends to be confined to specialised applications. It is used because it is sometimes cheaper or faster or indeed better, but these occasions tend to be rare.

The technique is often used for the analysis of one component of a mixture, especially when the compounds in the mixture are alike chemically or very similar in physical properties eg structural isomers. Here, analysis using uv/visible spectrometry is difficult because the spectra of the components will be nearly identical. Chromatographic analysis may be of limited use because separation, of say isomers, is difficult to achieve. We have already seem that the infrared spectra of isomers are usually very different in the fingerprint region and these bands can be used in the analysis. Another advantage of the technique is that it is non-destructive and requires little material.

In this Part of the Unit we shall first explore the relationship between the concentration of a solution and the amount of infrared radiation absorbed by it, and the special problems that the infrared technique imposes. Then we will test the theory to see if it always works or whether some solutions deviate from ideal behaviour. We shall then look at a few real applications followed by the impact of the recent introduction of the microcomputer (data stations) and how they have simplified the technique. We shall end by looking at industrial applications.

6.1. THE BEER–LAMBERT LAW

Lambert found in the eighteenth century that the amount of light transmitted by a solid sample was dependent on the thickness of the sample. This was extended to solutions during the following century by Beer. The resulting 'law' can be derived theoretically and applies to all electromagnetic radiation.

The statement above was intentionally 'woolly', and the word 'amount' does not convey anything about the relationship.

∏   Can you recall the mathematical equation that relates the light absorbed to the other variables mentioned above?

Well, it can be shown that the *absorbance* of a solution is *directly proportional* to the thickness and the concentration of the sample, or, in the form of a mathematical equation:

$$A = \epsilon c l \qquad (6.1)$$

where A is the absorbance of the solution, c is the concentration and l the thickness, or in infrared spectroscopy, the path-length of the sample. The constant of proportionality is usually given the symbol epsilon (ϵ), and referred to as molar absorptivity.

SAQ 6.1a	If the units used for concentration and path-length are mol dm^{-3} and mm respectively, what are the units of molar absorptivity?

SAQ 6.1a

We shall use m^2 mol^{-1} as the units for molar absorptivity throughout this Unit. When using values from the literature check the units, since path-lengths in centimetres may have been used. The units may then be dm^3 mol^{-1} cm^{-1}, and the value 10 times that when 'our' units are employed.

The absorbance can be shown to be equal to the difference between the logarithms of the intensity of the light entering the sample (I_o) and the intensity of the light transmitted (I) by the sample, ie

$$A = \log_{10}I_o - \log_{10}I = \log_{10}(I_o/I)$$

Absorbance is therefore dimensionless.

In this Unit we have usually used *% Transmittance* as the measure of the amount of light absorbed by a sample. What is the relationship between absorbance and transmittance? I will define transmittance and then ask you a question.

Transmittance,

$$T = -I/I_o \qquad (6.2)$$

and % Transmittance,

$$\%T = 100 \times T \qquad (6.3)$$

SAQ 6.1b	Try to write down the Beer–Lambert Law in a form that relates % transmittance to concentration and path-length. Is the constant of proportionality in this equation the same as in the equation involving absorbance?

Now that you have worked out the relationship between absorbance and % transmittance, let's see if we can get a feeling for the values that can be obtained for the absorbance of a solution.

First of all let us restate our final equation.

$$A = -\log_{10}(\%T/100) = \epsilon c l \qquad (6.4)$$

When using % transmittance values it is easy to relate and to understand the numbers. For example, 50%T means that half the light is transmitted and half is absorbed. 75%T means that three quarters of the light is transmitted and one quarter absorbed.

∏   What would be the absorbance of a solution which had a %
     transmittance of:

   (a) 100%   (b) 50%   (c) 10%   (d) 0%

Substituting the numbers above into Eq. 6.4, gives

$$(a)\ A = -\log_{10}(100/100) = -\log_{10}(1.0) = 0.0$$
$$(b)\ A = -\log_{10}(50/100)\ = -\log_{10}(0.5) = 0.303$$
$$(c)\ A = -\log_{10}(10/100)\ = -\log_{10}(0.1) = 1.0$$
$$(d)\ A = -\log_{10}(0/100)\ \ = -\log_{10}(0.0) = \text{infinity}$$

The important point to take away from these figures is that at an
absorbance of 1.0, 90% of the light is being absorbed, so the instrument's detector system does not have much radiation to work
with!

∏   What would be the % transmittance of a solution which had
     an absorbance of:

   (a) 0.2   (b) 0.5   (c) 1.0   (d) 3.0

To work out this series we need to transpose Eq. 6.4,

since $A = -\log_{10}(\%T/100)$

then $\%T = 100 \times 10^{-A}$

Hence,

(a) $\%T = 100 \times 10^{-0.2} = 63.1$

(b) $\%T = 100 \times 10^{-0.5} = 31.6$

(c) $\%T = 100 \times 10^{-1.0} = 10.0$

(d) $\%T = 100 \times 10^{-3.0} = 0.1$

You can now clearly see the exponential relationship from (c) and (*d*) above. It would make sense to try to work between absorbances of 0.2 and 0.5 since these correspond to 2/3 and 1/3 of the light being transmitted. At higher values of absorbance the detector system has little energy to work with and errors are introduced.

It is clearly important for you to be able to use the equations above in real analytical situations, so try the two questions below.

SAQ 6.1c

A compound has an absorption band at 830 cm^{-1} with a molar absorptivity of 2.0 m^2 mol^{-1}, calculate the range of concentrations that would give a satisfactory analytical peak using a path-length of 1.0 mm. (I mean a minimum absorbance of 0.2 and a maximum of 0.5.)

SAQ 6.1d

> A 1.0% w/v solution of hexan-1-ol has an absorbance of 0.37 at 3660 cm^{-1} in a 1.0 mm cell. Calculate its molar absorptivity at this frequency.

The Beer–Lambert Law tells us that a plot of absorbance against concentration should be linear with a gradient of ϵl and pass through the origin. In theory then, all we have to do to analyse a solution of unknown concentration is to prepare solutions of known concentration, choose a suitable absorption peak, measure the absorbance at this frequency, and plot the graph (a calibration graph).

We can now read the concentration of the compound in solution given its absorbance. Nothing could be easier. There are lots of hidden pitfalls however. Let's take each step in turn and try to isolate some of them.

1. *Prepare Solutions of Known Concentration.*

The concentrations have to give us sensible absorbance values. We therefore have to know the molar absorptivity in order to make up the standard solutions. There will therefore be some trial and error here.

2. *Choose a Suitable Absorption Peak.*

We would like the technique to be as sensitive as possible, so we should choose an intense peak. However, from the spectra you have

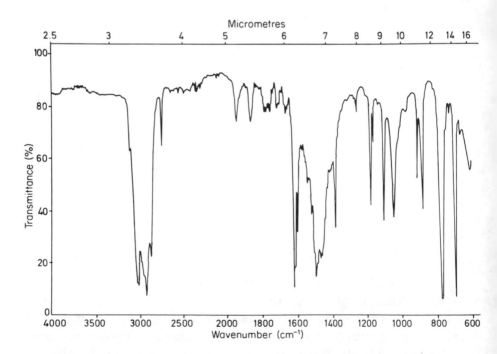

Fig. 6.1a. *Typical presentation of an infrared spectrum*

already seen it is obvious that infrared spectra have many, often overlapping, peaks. We try to find a peak isolated from others with a high molar absorptivity. Look at Fig. 6.1a. Which peak would you choose?

Very difficult, most peaks have shoulders or are very close to others. I would choose either the peak at 760 or 680 cm^{-1}.

A further problem that sometimes arises especially in spectra of solid samples is asymmetric bands. Here peak height cannot be used because the baseline will vary from sample to sample, and peak area measurements are used instead. This is now easily done automatically by modern microcomputer systems.

3. *Measure the Absorbance.*

Real problems here!

Most routine infrared spectrometers output spectra linear in % transmittance. We may therefore have to convert all our readings from transmittance to absorbance.

∏   I can think of three ways to do this (there could be more), can you write down two?

(*a*) I suppose we could use the % transmittance form of the Beer–Lambert Law we derived above in SAQ 6.1b, ie,

$$-\log_{10}(\%T/100) \;=\; \epsilon cl \qquad (6.4)$$

That was one of my not-so-good ideas, since all we have done is to convert transmittance to absorbance and the idea was to try to avoid that! However you could solve the problem by using log/linear graph paper. Here you would plot %T/100 against concentration, remembering that the gradient is now $-\epsilon l$.

(*b*) Much more cunning – examine Fig. 6.1a again and the Fig. 6.1b below.

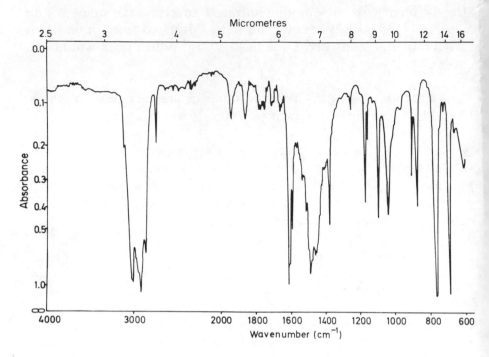

Fig 6.1b. *An infrared spectrum presented with a non-linear absorbance scale*

Do you see any difference between them? The chart paper is different. The spectrum in Fig. 6.1b is logarithmic in absorbance, instead of linear in % transmittance. This paper is easily available and carries out the calculations for you. It is of course difficult to read for low transmittance values (eg the peaks I have chosen), but that is a warning that your readings will not be all that accurate.

(*c*) Buy an instrument which will record in absorbance; it may be difficult to convince an employer.

SAQ 6.1e Measure the absorbance of the band marked A on Fig. 6.1c. Note the logarithmic scale. You shall have to interpolate carefully.

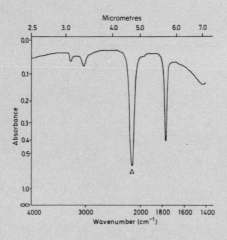

Fig. 6.1c

SAQ 6.1e

Did you agree? If you did then you must be an old hand at this game. The problem is where to measure from. The zero absorbance line on the chart paper? This is not suitable since it is not a reproducible base line. We need a zero line that will not vary from day to day, since it will depend on instrument settings, the cell used etc.

It is usual in quantitative infrared spectroscopy to use a base line joining the points of lowest absorbance on the peak, preferably in reproducibly flat parts of the absorption line (see Fig. 6.1d below).

The absorbance difference between the base-line and the top of the band is then used.

4. *Plot the Calibration Graph.*

This may be where your problems really begin. The graph may not be linear. Can you think of any reasons why this should be so? I will leave that one with you for the present and deal with it after this question.

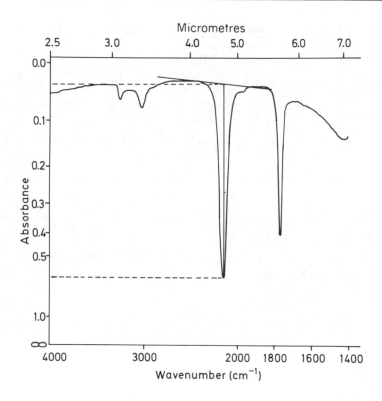

Fig. 6.1d.   *Base line construction*

<table>
<tr><td>SAQ 6.1f</td><td>I would like you to now go through the whole analysis process with real spectra. Fig. 6.1e (*i*)–(*v*) are infrared spectra of phenylacetylene (phenylethyne) in tetrachloromethane solution. Please determine the concentrations of phenylacetylene shown in the spectra in Fig. 6.1f, as % v/v.</td></tr>
</table>

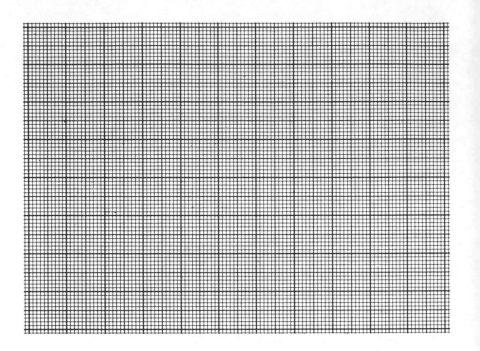

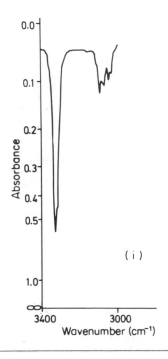

(i)

Wavenumber (cm⁻¹)

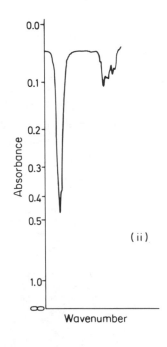

(ii)

Wavenumber

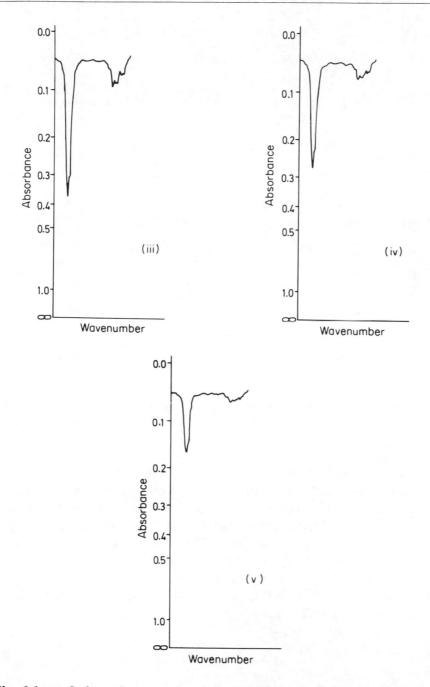

Fig. 6.1e. *Infrared spectrum of phenylethyne in CCl$_4$ (i) 5.00% (ii) 4.00% (iii) 3.00% (iv) 2.00% (v) 1.00% v/v*

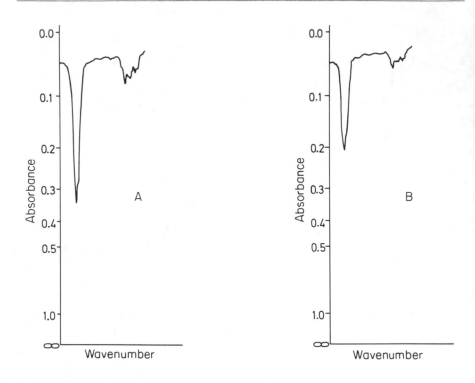

Fig. 6.1f. *Infrared spectrum of phenylethyne in CCl₄: Unknown concentrations A and B*

You have already come across one phenomenon which leads to deviation from the Beer–Lambert Law. You are not recommended to use the O—H stretching frequency as an analytical peak under normal conditions. You saw in Part Five that the intensity of the absorption bands in the 3000–3650 cm^{-1} region result from an equilibrium mixture of associated molecules. The relative amounts of these species are concentration dependent so any analytical absorption peak chosen in this region would give a non-linear plot of absorbance *vs* concentration.

SAQ 6.1g   Fig. 6.1g (*i*)–(*v*) are infrared spectra of propan-2-ol as 5, 4, 3 2 and 1% v/v solutions in CCl_4. Draw a Beer–Lambert plot from measurements of the O—H stretching band.

Is this graph linear? Could you use it as a calibration curve for the determination of propan-2-ol?

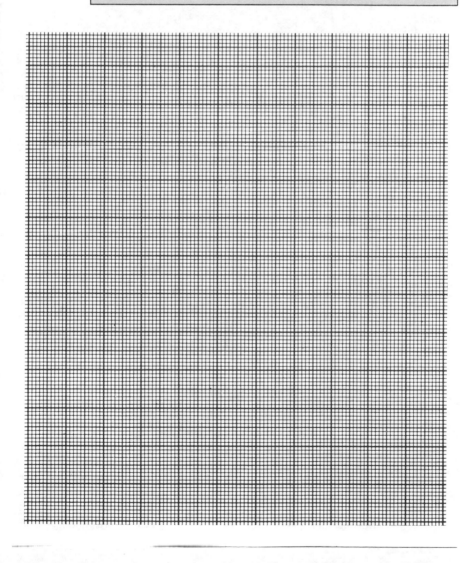

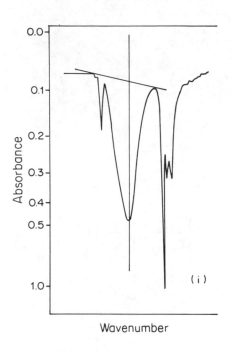

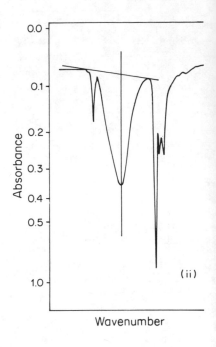

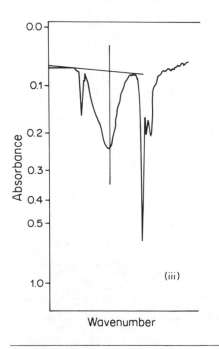

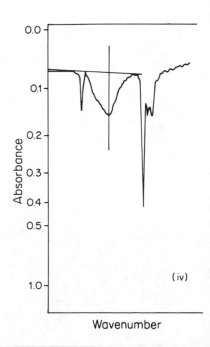

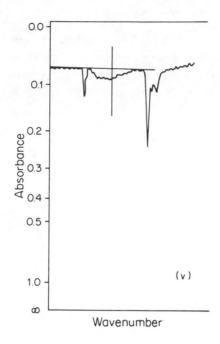

Fig. 6.1g. *Infrared spectrum of propan-2-ol in CCl_4 (i) 5.00%
(ii) 4.00% (iii) 3.00% (iv) 2.00% (v) 1.00% v/v*

The effect of hydrogen bonding is not confined to the O—H stretching frequency. We saw earlier that the C=O and N—O stretching
frequencies can also shift in hydrogen bonded materials. In fact any
type of molecular association will lead to non-ideal behaviour. In
analysis of mixtures, overlapping bands from the components can
lead to non-linearity. Instrumental factors can also be important. It
is found that if the slit width is greater than the peak width at half
height then the molar absorptivity of the band is lowered and since
the peak width varies with concentration, the calibration graph will
again be non-linear. Modern optical systems have to a large extent
eradicated this effect.

SAQ 6.1h
> It is possible to successfully use the O—H stretching absorption as an analytical peak. Suggest a way of doing this (your method will have to remove hydrogen bonding effects).

Examine the four spectra of propan-2-ol in pyridine, given in Fig. 6.1h. Indicate the differences in the O—H stretching band compared with the corresponding spectra recorded in tetrachloromethane, which you looked at a few minutes ago.

There is now no monomer band present even at low concentrations. This is because of the strong intermolecular hydrogen bond between the alcohol and the pyridine lone-pair electrons. Pyridine absorbs above 3000 cm^{-1} (the C—H stretching frequency), making it difficult to draw a baseline for the O—H stretching frequency. I have drawn two on the spectra and tabulated the absorbances below.

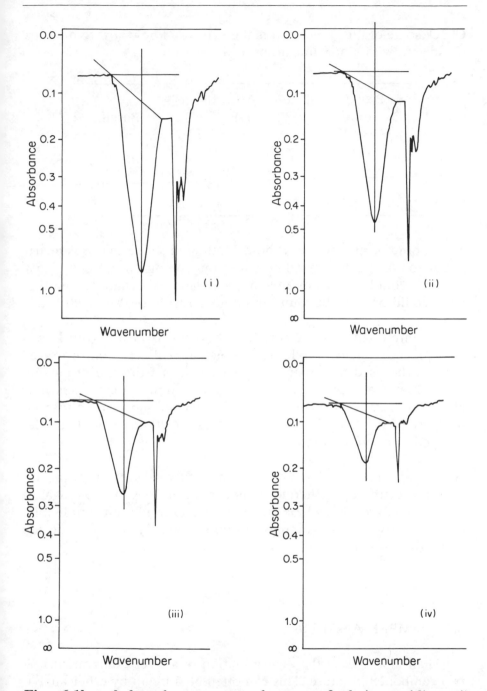

Fig. 6.1h. *Infrared spectrum of propan-2-ol in pyridine (i) 5.00% (ii) 2.50% (iii) 1.25% (iv) 0.625% v/v*

Check some of the values and make sure you know how to measure up spectra with a non-linear absorbance scale.

% Propan-2-ol in Pyridine (%v/v)	Absorbance Baseline 1	Baseline 2
5.00	0.677	0.723
2.50	0.371	0.403
1.25	0.182	0.205
0.625	0.102	0.118

You should be able to see without plotting this data that is is nearly linear for the second baseline (the non-sloping one), below 2.5% v/v. This could be used as an analytical band. Alternatively another band could perhaps be found in another region of the spectrum.

The example you worked through above in SAQ 6.1f (the Beer's Law plot for phenylacetylene) was not a 'real' example. Infrared tends to be used for mixtures. We have seen from the discussion above that the presence of the second or more compounds may affect the intensity of bands of the first compound leading to non-linearity of the Beer–Lambert plot. Many methods have been developed to minimise these effects.

Any newly developed method for quantitative analysis using infrared spectroscopy, therefore, needs careful evaluation against known standards. Let's look at some 'real' examples on simple mixtures then some more complex examples where more sophisticated methods are required.

6.2. SIMPLE ANALYSIS

We shall first look at the determination of a trace component in a two component mixture. This can often be difficult by other analytical methods. The success of this method depends on the presence of an intense band due to the impurity.

Commercial propan-2-ol often contains traces of acetone formed by oxidation:

$$
\begin{array}{c}
\text{OH} \\
| \qquad \text{[O]} \\
\text{CH}_3\text{CHCH}_3 \longrightarrow \text{CH}_3\text{COCH}_3
\end{array}
$$

Fig. 6.2a is a spectrum of pure acetone while Fig. 6.2b is a spectrum of pure propan-2-ol. Study these and choose an analytical peak.

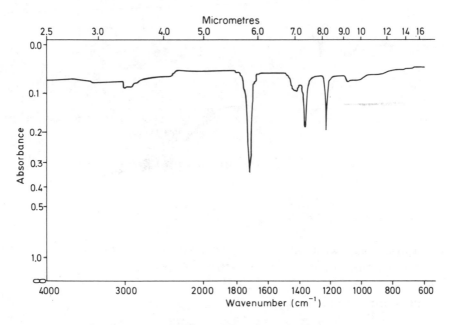

Fig. 6.2a. *Infrared spectrum of acetone*

There are two ways to approach this. We can either use a peak due to acetone or one from propan-2-ol. In this case, where we are interested in the low concentration of acetone, there will be little change in peak intensities from the propan-2-ol as its concentration varies from say 95 to 100%. Clearly we need a suitable peak from acetone and it is a matter of choosing the one which gives the best accuracy.

The chosen peak should:

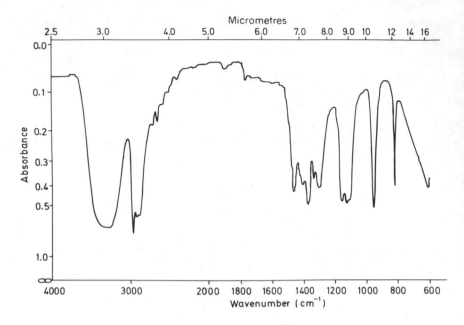

Fig. 6.2b. *Infrared spectrum of propan-2-ol*

— have a high molar absorptivity,

— not overlap with or be close in frequency to other peaks from other components in the mixture or solvent,

— be symmetrical,

— ideally have an absorbance between 0.2 and 0.5 in the solutions to be used,

— give a linear calibration plot of absorbance versus concentration.

Inaccuracy can easily arise from other sources, such as badly calibrated cells and incorrect spectrometer settings such as slit widths or scanning speeds. You will recall that we discussed the maintenance and calibration of cells in Part 3.

Which peak did you decide on? I would choose the C=O stretching absorption of acetone, because it is an intense peak and is in a region where no absorptions occur from the other component of the mixture.

The next step in the analysis is to draw a calibration plot of absorbance against concentration.

SAQ 6.2a
Draw a plot of absorbance against concentration from the data below and calculate the molar absorptivity in units of $m^2 \, mol^{-1}$.

Acetone in CCl_4 (%v/v)	% Transmittance at 1719 cm^{-1}
0.25	65.6
0.50	48.5
1.00	26.9
1.50	16.0
2.00	10.0

The transmittance values given were read straight from the spectrum. The baseline had a transmittance of 86% at 1719 cm^{-1}.

The density of acetone is 0.790 $g.cm^{-3}$.

The path-length was 0.1 mm.

Warning – this may take you half an hour, so this may be a good time to break.

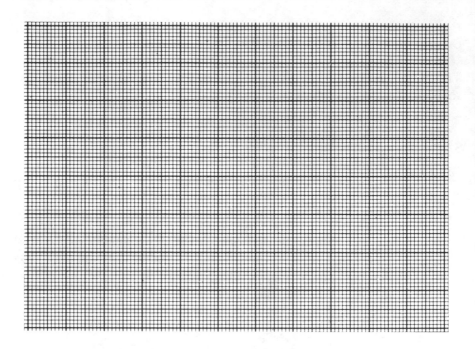

I hope you had some success with that. I'm also sure it took you a long time, but it is important to work through a real example – let's go on.

The next stage in the analysis is to look at the spectrum of a sample of propan-2-ol and determine the amount of acetone present.

SAQ 6.2b	The infrared spectrum of a 10% v/v solution of commercial propan-2-ol in CCl_4 in a 0.1 mm path length cell is shown in Fig. 6.2e.

(i) Determine the concentration $(mol\,dm^{-3})$ of acetone in this solution using the calibration curve plotted in SAQ 6.2a.

(ii) perhaps more important, calculate the % acetone in the propan-2-ol. $\longrightarrow$

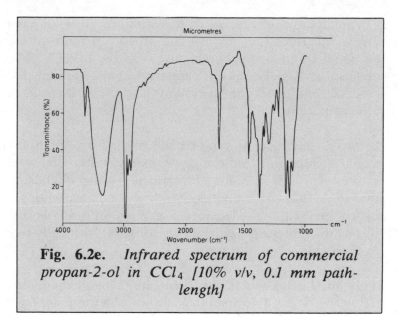

Fig. 6.2e. *Infrared spectrum of commercial propan-2-ol in CCl₄ [10% v/v, 0.1 mm pathlength]*

Chromatographic analysis of this mixture gives an answer close to 7% v/v acetone. Our error is due to background differences at the carbonyl frequency between pure acetone in CCl_4. The propan-2-ol absorbs slightly at this frequency. This problem leads to large errors in multicomponent systems and methods have been devised to eliminate it.

Infrared spectroscopy can be used to measure the number of functional groups in a molecule, say the number of $-OH$ or $-NH_2$ groups. This has been used in natural product chemistry. It is found that the molar absorptivity of the bands corresponding to the group is proportional to the number of groups. In other words each group has its own intensity, which does not vary drastically from molecule to molecule.

This has been used to measure chain length in hydrocarbons by using the $C-H$ deformation bands of the methylene group at 1467 and 1305 cm^{-1} and the number of methyl groups in polythene.

6.2.1. Analysis of Solid Samples

Solid mixtures can also be analysed. They are more susceptible to errors because of scattering and solid state anomalies.

These analyses are usually carried out on KBr discs or in mulls. The problem here is the difficulty in measuring the path length. This measurement becomes unnecessary, however, when an internal standard is added to the samples and to the calibration standards. Addition of a constant known amount of an internal standard is made to all samples and calibration standards.

The calibration curve is then obtained by plotting the ratio of absorbance of the analyte to that of the internal standard, against concentration of the analyte, the samples being contained in a mulling agent or KBr. The absorbance of the internal standard varies linearly with sample thickness and thus compensates for it. The discs or mulls must be made under exactly the same conditions to avoid intensity changes or shifts in band positions because of solid state anomalies.

The standard must be carefully chosen and it should ideally,

— have a simple spectrum with very few bands,

— be stable to heat and not pick up moisture,

— be easily reduced to a particle size less than the incident radiation without lattice deformation,

— non-toxic, giving clear discs in a short time,

— readily available in the pure state.

Standards used have included calcium carbonate, sodium azide, naphthalene and lead thiocyanate.

6.3. MULTICOMPONENT ANALYSIS

The analysis of a component in a complex mixture presents special problems. You saw above that the background absorption from propan-2-ol in the acetone/propan-2-ol mixture led to an error in the analysis. In more complex mixtures this is likely to increase. Let's look at two examples, each illustrating a different method that has been devised to minimise these effects.

(a) Analysis of a Four Component Mixture

Commercial xylene is a mixture of isomers, and on first thinking about this, I would expect 1,2-dimethylbenzene (*o*-xylene), 1,3-dimethylbenzene (*m*-xylene) and 1,4-dimethylbenzene (*p*-xylene).

There is also another isomer present, but I will leave its structure until later. We shall try to devise a method for the analysis of all four components. The spectra of the three pure xylenes, Figs 6.3a–6.3c below, all show strong bands in the 650–800 cm^{-1} region. It is found that cyclohexane has very low absorbance in this region and is therefore a suitable solvent for the analysis. Analysis of mixtures of this sort can be achieved by successive approximations to the actual mixture. First the concentrations of the components are approximated from standards, a test solution is made up at these concentrations and compared with the unknown. By repeating this process a spectrum can be eventually got which exactly matches the unknown. Minor components can then be estimated by difference spectroscopy. By this I mean placing the made up mixture in the back beam of the instrument and the analytical solution in the front beam and recording the spectrum, the spectrometer will output a difference spectrum.

SAQ 6.3a

Measure the absorbance of the xylenes (*o m* and *p*) at 740, 770 and 800 cm^{-1} in the spectra given in Fig. 6.3a, b and c.

The infrared spectrum of a commercial sample of xylene is given in Fig. 6.3d. Estimate the concentrations of the three isomers in this sample.

All the spectra were recorded using the same cell which had a path-length of 0.1 mm.

Note that the solvent used in all cases was cyclohexane but that the concentrations of the four samples was not always the same:

o-xylene 1% v/v
m-xylene 2% v/v
p-xylene 2% v/v
commercial xylene 5% v/v

There are a few short-cuts that I hope you will spot. If not they are in the answer.

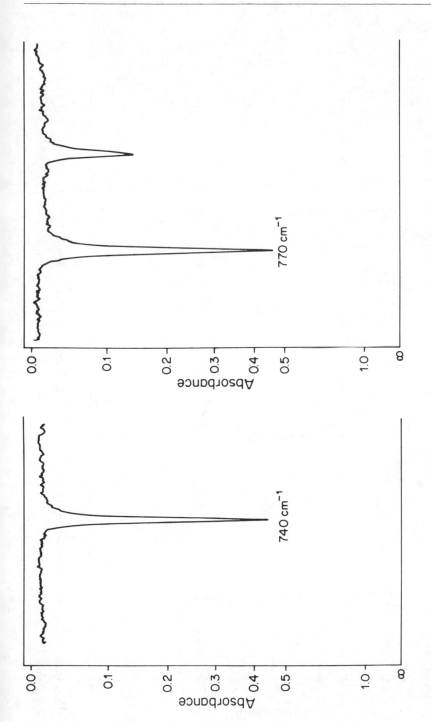

Fig. 6.3b.   *Infrared spectrum of m-xylene in cyclohexane. [2% v/v, 0.1 mm path-length]*

Fig. 6.3a.   *Infrared spectrum of o-xylene in cyclohexane. [1% v/v, 0.1 mm path-length]*

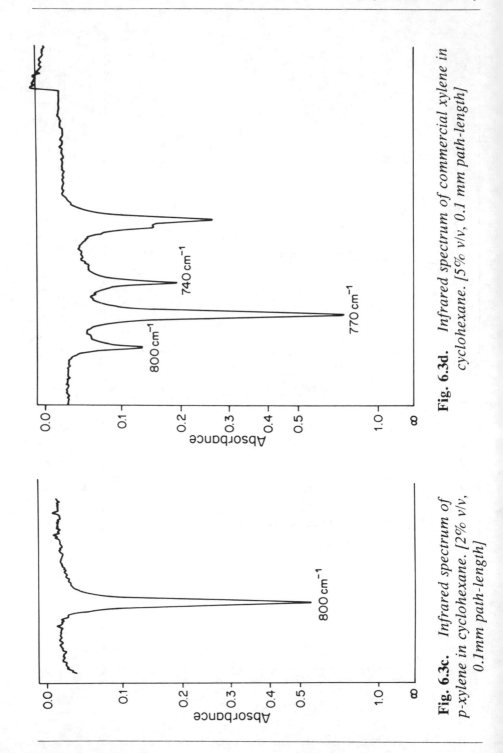

Fig. 6.3c. *Infrared spectrum of p-xylene in cyclohexane. [2% v/v, 0.1mm path-length]*

Fig. 6.3d. *Infrared spectrum of commercial xylene in cyclohexane. [5% v/v, 0.1 mm path-length]*

SAQ 6.3a

I hope you worked that out for yourself and found that the three components did not add up to 5%.

SAQ 6.3b

Can you think of three reasons for the discrepancy between the calculated values of the concentration of the isomers and the total concentration?

SAQ 6.3b

The next stage is to make up a solution at these calculated concentrations, ie,

o 0.364%,   *m* 3.076%   and   *p* 0.389%,   all v/v.

Fig. 6.3e is a spectrum of this solution. It is now possible to use this spectrum to take overlapping bands into account.

∏   Do we need to change the concentrations we calculated above?

Fairly obviously, but it is possible to get better matching spectra; look at Fig. 6.3f, this matches fairly closely with the commercial sample in Fig. 6.3d.

The concentrations used were:

o 0.46% v/v,   *m* 3.26% v/v,   *p* 0.55% v/v,

This totals to only 4.27% v/v, so there must be another component.

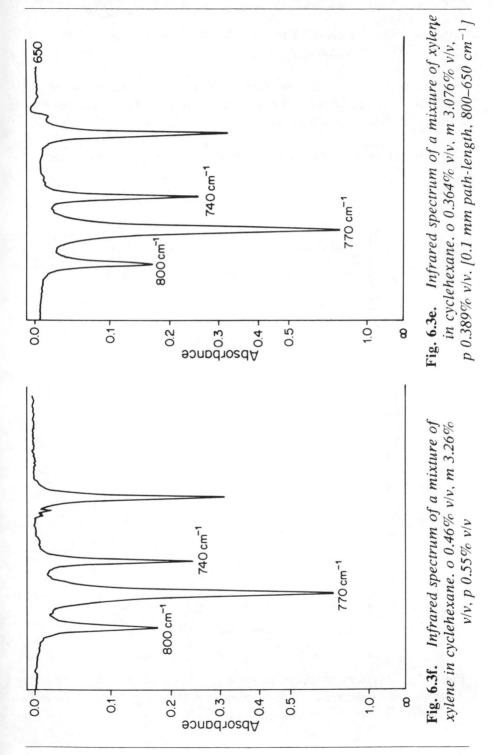

Fig. 6.3e. *Infrared spectrum of a mixture of xylene in cyclehexane. o 0.364% v/v, m 3.076% v/v, p 0.389% v/v. [0.1 mm path-length, 800–650 cm⁻¹]*

Fig. 6.3f. *Infrared spectrum of a mixture of xylene in cyclehexane. o 0.46% v/v, m 3.26% v/v, p 0.55% v/v*

Note the shoulder on the band at lowest frequency in the spectrum of the commercial sample.

If this latest mixture is now placed in the back beam of the instrument and the commercial xylene sample in the front beam, a difference spectrum will result.

The difference spectrum, 850 to 650 cm^{-1} is shown in Fig. 6.3g.

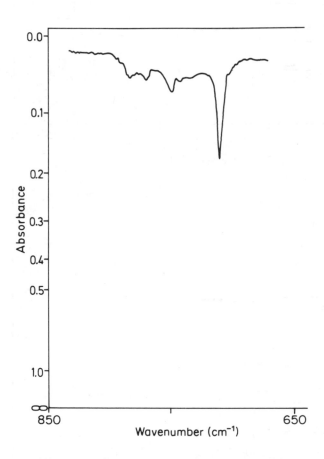

Fig. 6.3g *Difference spectrum for a synthetic mixture of xylene isomers and a commercial xylene sample [0.1 mm path-lengths, 850– 650 cm^{-1}]*

By comparison with library spectra, it can be shown that this component is ethylbenzene. Look at Fig. 6.3h, below. This is a spectrum of pure ethylbenzene, – convinced?

Ethylbenzene can now be determined using the method above.

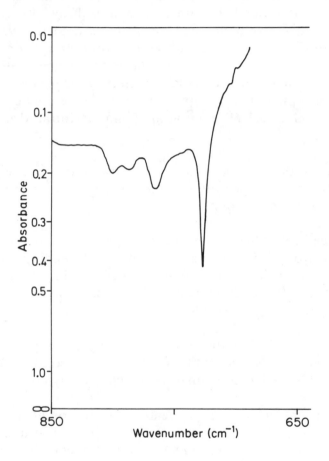

Fig. 6.3h. *Infrared spectrum of ethylbenzene*

(*b*) *Determination of Trans Unsaturation in Fats*

Naturally occurring vegetable fats and oils are a mixture of triglycerides of general formula,

$$CH_2O.CO.R_1$$
$$|$$
$$C\ HO.CO.R_2$$
$$|$$
$$CH_2O.CO.R_3$$

This is a tri-ester of glycerol. The acids are long chain and can be saturated or unsaturated. The unsaturation is always present as the *cis* isomer. These *cis* bonds can be isomerised to the *trans* configuration during extraction or subsequent processing. For example oxidation or partial hydrogenation can lead to isomerisation. It is commercially important (for labelling etc.) to determine this *trans* content. It is difficult to separate *cis* from *trans* isomers using other techniques such as gas chromatography and an infrared method is commonly used.

The beauty of the method described below is that no weighing is required nor are any accurate volumes needed and it can therefore be used for very small samples.

You will see in Part Seven that the configuration of alkenes can be determined from the frequency of their $C-H$ bending vibrations. *Cis* isomers absorb between 700 and 840 cm^{-1} while *trans* isomers absorb between 930 and 1000 cm^{-1}. This band can be used as the basis for the analytical method.

It is normal to hydrolyse the triglyceride mixture to glycerol and a mixture of fatty acids, and the latter then converted to their methyl esters. It is this methyl ester mixture which is analysed.

I give the formulae of methyl oleate, the *cis* isomer and methyl elaidate, the *trans* isomer, below. Both geometrical isomers absorb strongly at 1163 cm^{-1}, the $C-O$ stretching frequency from the ester group.

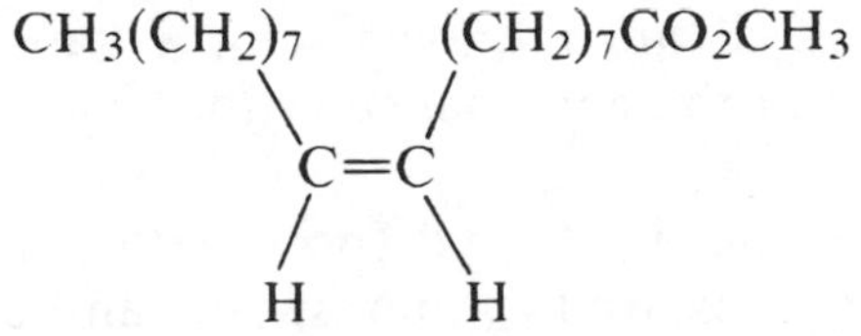

$$CH_3(CH_2)_7 \qquad (CH_2)_7CO_2CH_3$$

$$C=C$$

$$H \qquad H$$

Methyl oleate
(*cis*)

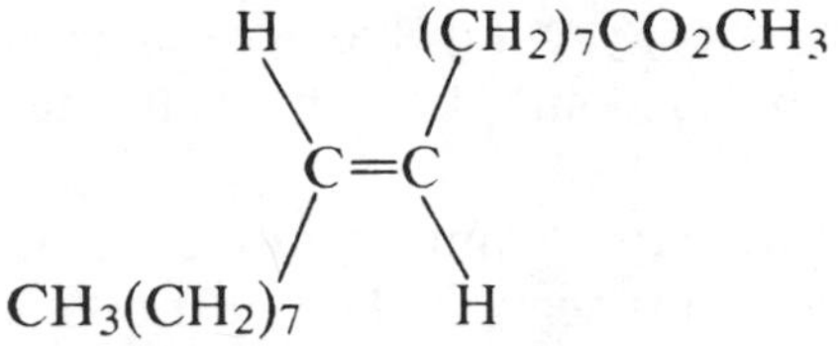

$$H \qquad (CH_2)_7CO_2CH_3$$

$$C=C$$

$$CH_3(CH_2)_7 \qquad H$$

Methyl elaidate
(*trans*)

Cis isomers absorb weakly at 965 cm^{-1} while *trans* isomers absorb strongly at this frequency. It can be shown that:

$$\%\,trans = K(A_{965}/A_{1163}) - f$$

where K and f are constants, and A_{965} and A_{1163} are the absorbances of the solution at these frequencies respectively.

Note that no concentration or path-length term appears in this equation.

SAQ 6.3c How would you go about finding the constants in the equation above, ie

$$\%\,trans = K(A_{965}/A_{1163}) - f$$

Once K and f have been evaluated an unknown sample can be analysed simply by measurement of the absorbance ratio.

I am now going to move on to some light relief (no questions) to end this section. The next two sections are for interest only and can be read at leisure, you no longer need that pen, paper and graph paper!

6.4.   USE OF DEDICATED MICROCOMPUTERS

The last few years has seen the introduction of cheap microcomputers using floppy disc and/or Winchester disc storage. Instrument manufacturers have quickly introduced interfaced 'Data-Stations' for their instruments. These, however, tend to be as expensive as the instrument, mainly due to the cost of software development and the relatively small market.

This section deals with the advantages that these microcomputers have brought to infrared spectroscopy.

These computers can perform the following functions:

(*a*) Control the instrument, eg, set scan speeds, slit widths, scanning limits, start and stop scanning etc.

(*b*) Read spectra into computer memory from the instrument as the spectrum is scanned. This means that the spectrum is digitised. This digitised spectrum can then be saved permanently on floppy disc for future use, comparison etc.

(*c*) Read stored spectra from floppy disc into memory and display on a monitor.

(*d*) Plot these spectra either on the instrument's recorder or on a separate plotter.

(*e*) Manipulate spectra, eg add and subtract spectra, expand areas of interest, reduce noise in spectra and so on.

(*f*) Scan spectra continuously and average or add the result in computer memory, thus reducing noise in very dilute solutions.

(*g*) Run complex analyses automatically by following a set of pre-programmed commands.

(*h*) Act as a normal microcomputer. It can thus be used as a word-processor or to run programs written in *Basic*, *Pascal* or other available languages.

In this section we shall concentrate on (*e*) above and illustrate the power of an interfaced microcomputer.

6.4.1. Manipulation of Spectra

(*a*) Subtraction of spectra can be used to eliminate solvent peaks. This is especially useful for mulls, when the spectrum of the mulling agent can be subtracted, giving the spectrum of the solid only. Fig. 6.4a is a nujol mull of potassium benzoate,

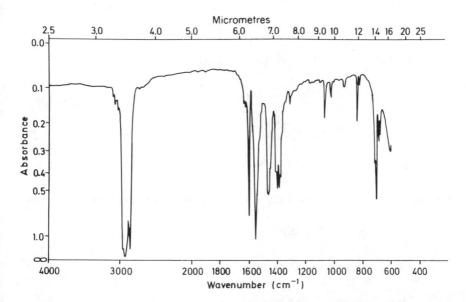

Fig. 6.4a. *Infrared spectrum of potassium benzoate [nujol mull]*

Fig. 6.4b is the spectrum of nujol and the difference spectrum is shown in Fig. 6.4c.

A data station can also be used to determine the structure of impurities and if a sample of this compound is available, mixtures of starting material and product can be 'separated' by subtraction of the spectra of the mixture and impurities. In this way the analysis of commercial xylene outlined above could be automated by a series of subtractions. This is achieved by subtracting a fraction of o-, then m- and finally p-xylene from the commercial xylene spectrum to null the corresponding absorptions; these factors can be read from the screen.

I have done this with the same spectra used above and produced the following result:

Compound	Subtraction Factor	Concentration (% v/v)
o-xylene	0.35	1.0
m-xylene	1.45	2.0
p-xylene	0.29	2.0

These results therefore give the following composition:

o   -xylene   0.35% v/v

m   -xylene   2.90% v/v

p   -xylene   0.58% v/v

The first method we used gave the following compositions;

o   -xylene   0.46% v/v

m   -xylene   3.26% v/v

p   -xylene   0.55% v/v

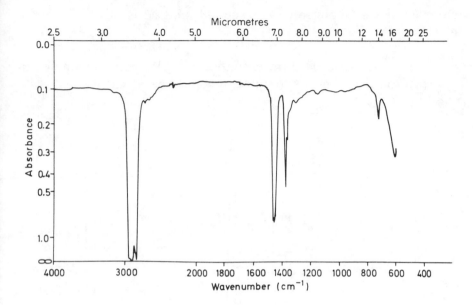

Fig. 6.4b. *Infrared spectrum of nujol*

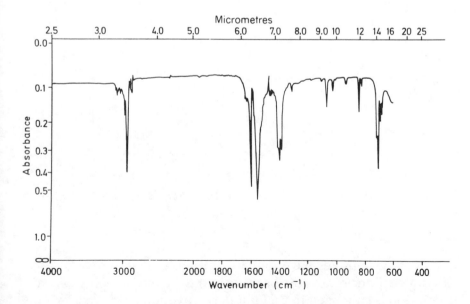

Fig. 6.4c. *Infrared spectrum of potassium benzoate after subtraction of nujol bands*

which compare favourably. The second method is of course much faster.

(*b*) Spectra can also be differentiated. Fig. 6.4d shows a single absorption peak and its first and second derivative.

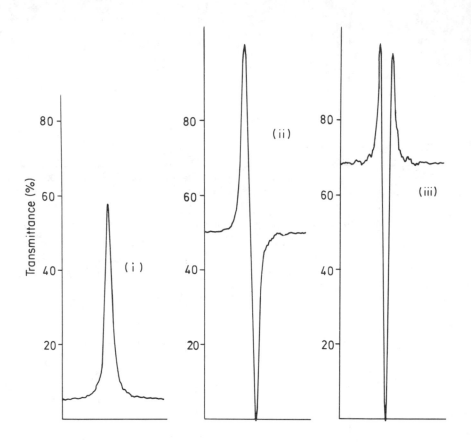

Fig. 6.4d. *An absorption peak (i), the first derivative (ii) and the second derivative (iii)*

Derivative techniques have long been used in quantitative ultraviolet/visible spectroscopy, the benefits of the technique being two-fold. Resolution is enhanced in the first derivative since we are now looking at changes in gradient. The second derivative gives a negative peak for each band and shoulder in the absorption spectrum.

The advantages of derivatisation are more readily appreciated for more complex spectra and Fig. 6.4e shows how differentiation can be used to resolve and locate peaks in an envelope. Note that sharp bands are enhanced at the expense of broad ones and that this may allow selection of a peak even when there is a broad band beneath. This latter point is clearly demonstrated in Fig. 6.4f.

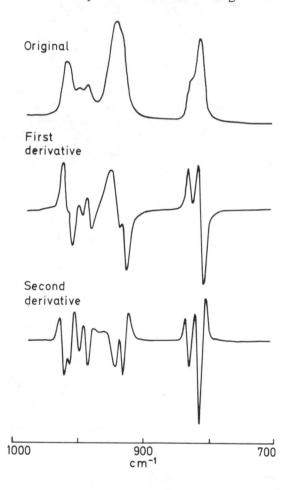

Fig. 6.4e. *An Infrared spectrum recorded in absorption, first derivative and second derivative modes. © Perkin–Elmer, 1984. Reproduced by permission of Perkin–Elmer Ltd.*

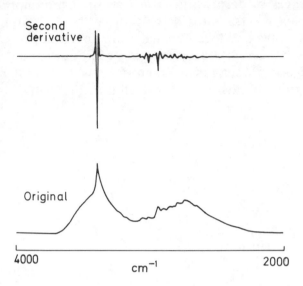

Fig. 6.4f.   *The resolution of a sharp peak from a broad background using second derivative spectroscopy. © Perkin–Elmer, 1984. Reproduced by permission of Perkin–Elmer Ltd.*

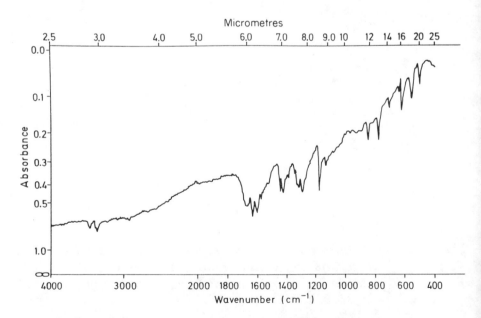

Fig. 6.4g.   *Infrared spectrum of a sample prepared as KBr disc*

(*c*) Spectra can be readily manipulated to aid quantitative measurement. Parts of the spectrum can be selected and the ordinate and abscissa expanded. Conversion from % transmittance to absorbance is very rapid and allows peak height (absorbance) to be displayed automatically. Areas can also be calculated automatically, the operator having first set a baseline.

(*d*) Noise can be diminished by 'smoothing' (at the expense of resolution) and scattering from mulls and discs compensated to restore horizontal baselines.

We have recorded an infrared spectrum of a KBr disc, restored the baseline and finally smoothed the spectrum in Fig. 6.4g, Fig. 6.4h and Fig. 6.4i.

(*e*) It is also possible to preprogram a whole analysis, by issuing commands to the operator and automatically performing the analysis. This allows unskilled staff to use the technique in industry, in many instances on the plant. The main advantage is

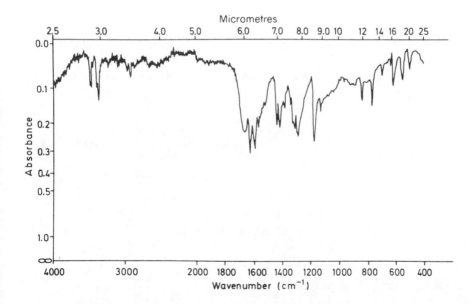

Fig. 6.4h. *Infrared spectrum displayed after computer correction of a sloping baseline*

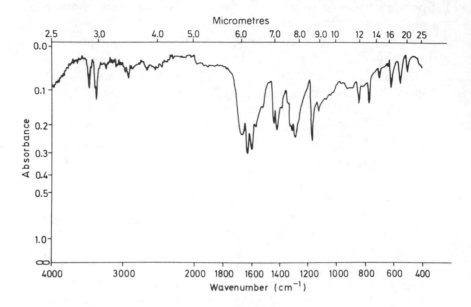

Fig. 6.4i. *Infrared spectrum displayed after computer 'smoothing' to suppress noise*

the time saved, since the sample may not need laboratory analysis. When monitoring continuous processes, cost effectiveness can often be improved dramatically when analytical results are produced very rapidly.

6.5. USE OF INFRARED MONITORING IN CHEMICAL PLANTS

Infrared radiation is ideal for on-line analysis in industrial environments. It is now also widely used to monitor atmospheric pollution. I shall take these two applications in turn.

6.5.1. On-line Analysis

Infrared monitors can be built into the sample stream and used to monitor a band in the required product.

This does not mean that we only have to transport our laboratory instrument into the plant. The differences in requirements between the two uses are enormous. Let's summarise the requirements of the two cases.

Laboratory

(*a*) A versatile instrument, capable of being used for many different applications.

(*b*) Large frequency range.

(*c*) Housed in a fairly 'clean' non-corrosive atmosphere.

(*d*) Will be treated (hopefully) carefully and therefore need not be too robust.

Plant

(*a*) Needs to be rugged in construction.

(*b*) May be in a highly corrosive environment and may need to be purged with dry air or nitrogen.

(*c*) Needs to be capable of continuous, reliable analysis, 24 hours a day.

(*d*) Usually dedicated to one task, may only need a single wavelength source, if this is possible.

These are quite different animals!

The plant analyser may be cheaper, mainly because it will be dedicated and needs little versatility.

The most important point in plant analysis is the sample and whether it is representative, since if it isn't then the whole exercise is useless. Gases are often analysed so here pressure must be kept constant. The gas may cool and condense on windows, and the windows themselves have to be chosen carefully (because of sample or environment) and may therefore dictate the wavelength chosen for the analysis. The windows must also be easily accessible for maintenance. The system must also be easily sampled to allow calibration.

Many analysers depend on filtering out interfering wavelengths. If a mixture of three gases, A, B and C are to be monitored for the concentration of A, the beam can be passed through a sample of B and C thus removing the interfering wavelengths, then passed through the sample. Tunable single wavelength laser sources are now available which can be used. Other instruments use two wavelengths and use the ratio method of analysis outlined above. Many different instruments are commercially available.

6.5.2.   Pollution Monitoring

The routine monitoring of vapour levels of toxic chemicals in the work environment with the added back-up of recent legislation in many counties has led to the development of the infrared technique for the automatic quantitative analysis of noxious chemicals in air.

The technique is based on long path-length cells with a pump to circulate the air sample through the instrument. Simple instruments are dedicated to a single contaminant in air, employing a filter system for wavelength selection. These are commonly used for SO_2, HCN, phosgene, NH_3, H_2S, formaldehyde and HCl. More complex instruments are tunable for a particular analytical band and usually give a direct readout in ppm or % v/v of the pollutant. The path-length of the cell can also be varied, from less than a metre to 20 metres.

The instruments are rugged, easily portable (battery operated) and can be used to monitor any chemical with an active ir band. Interference from other pollutants can be minimised by the correct choice of analytical frequency. Accuracy is as good as 2%. A table is given below for a few common pollutants with the recommended frequencies. The minimum levels quoted are for a 20 m cell.

Compound	Analytical Frequency (cm^{-1})	Pathlength (m) Recommended	Min. Detectable Conc. (ppm)
CH_3CN	1042	20	5.0
NH_3	962	20	0.2
C_6H_6	672	20	0.3
$CH_3CH_2COCH_3$	1176	20	0.15
CCl_4	794	20	0.06
CS_2	2203	20	0.5
CO_2	2353	0.75	0.5
CO	2169	20	0.2
CCl_2F_2 (Freon 12)	1099	0.75	0.02
CBr_2F_2	1087	2.25	0.02
SO_2	1163	20	0.5
HCN	3290	20	0.4
CH_3NH_2	2941	20	0.1
$CH_2{=}CHCl$	917	20	0.3

Typical applications are monitoring of formaldehyde in plastic and resin manufacture, anaesthetics in operating theatres, degreasing solvents in a wide variety of industries and carbon monoxide in garages, I'm sure you can think of many more.

Summary

This Part of the Unit began with a revision of the Beer–Lambert Law, showing how the intensity of an absorption band is related to the amount of analyte present. This was then used in a simple analysis.

Techniques for dealing with the problem of background absorption were presented followed by treatment of the analysis of multi-component mixtures.

The use of microcomputers in analytical work was discussed, and the Part concluded with a brief introduction to the industrial use of on-line infrared analysis.

Objectives

On completion of Part Six you should be able to:

- state and use the Beer–Lambert Law in quantitative analysis;

- convert transmittance values to the corresponding absorbance values;

- appreciate that only a definite range of absorbance and transmittance can be used to give the greatest accuracy in quantitative analysis;

- choose a suitable analytical band and state reasons for the choice;

- plot absorbance against concentration calibration graphs and hence analyse simple mixtures;

- describe a variety of analytical techniques used for compensating for background absorption and overlapping peaks;

- describe the advantages of a dedicated microcomputer linked to an infrared spectrometer;

- compare laboratory and industrial analytical instrumentation.

7. Spectrum Interpretation

Well, you have reached the last (and most important) Part of this Unit.

The most important routine use of infrared spectroscopy is in monitoring reactions, that is, an answer to the question, *has my reaction given the desired product*? By the use of the techniques in the previous Part it can also tell you how far it has gone. It is extensively used however, to start to answer the question, *What is the structure of this unknown material*? You will see below that infrared can seldom answer this question completely so other techniques are used in conjunction, usually mass spectrometry and nmr spectroscopy.

Infrared is however a factor of ten cheaper to buy and cheap to run in comparison with ms and nmr, hence its widespread use in even the smallest industrial laboratory. Modern computer techniques have given new scope to ir, allowing easy spectrum subtraction to identify trace components, while modern Fourier instruments are fast enough to be coupled directly to glc and hplc instruments and give spectra of each fraction eluted from the columns. These techniques are likely to come into widespread use in the next few years.

You are therefore studying a developing analytical method and one in which a lot has still to be learnt and applied to new problems.

You have to walk before you can run, however, so I shall devote most of this Part to the interpretation of the spectra of fairly simple molecules.

You have already covered the reasons why some absorptions are more useful in this context than others (Part Four). Here we shall examine methods of interpretation of infrared spectra. This can be treated purely empirically using correlation charts (which you will find in Appendix A) but the knowledge you gained from Part Four should enable you to;

(*a*) understand why particular groups absorb at the frequencies they do,

(*b*) understand the reasons for the small differences in absorption frequencies of a particular group,

(*c*) understand why some groups absorb within a narrow frequency, range while others do not and are therefore fairly useless as structural indicators.

In this Part we shall be looking at many spectra to practice spectral interpretation. As a result it probably looks a little long to you! I can only say that there is no substitute for examining lots of spectra, and there are lots contained in this Part of the Unit.

First we shall consider the normally studied infrared region in four sections and stress the bands useful for structure determination. Then we shall consider one important stretching frequency in detail, trying to rationalise the differences in frequency of absorption with changes in neighbouring atoms. Then we will look at the common organic compounds by chemical class and the bands expected in their spectra.

I shall end with a series of interpretation problems and give you some to dip into later (at your leisure!). So don't feel you have to master this lot at one sitting, there is more work here than in some whole Units! Don't try to commit most of it to memory. Familiarity and practice in the technique 'locks in' the important wavenumber ranges, so work through the examples, then try to apply your knowledge to other samples as you come across them later at work (or play). You may find that you do not need all the detail in this Part of the Unit. If you have a mass spectrometer or an nmr instrument available, then you won't, so there are various ways you can work through it and this can only be gauged by you.

7.1. EMPIRICAL USE OF GROUP FREQUENCIES

We saw in Part One that the frequency of absorption by a particular vibrational mode of a molecule was determined by,

— the bond strength(s), or more correctly stiffness,

— the masses of the atoms concerned.

We also saw that certain vibrations were confined to one bond, while others could be referred to as coupled and involved many atoms. We saw that the second type varied in frequency with small changes in molecular structure, while the former type could be relied on to appear in the spectrum at the same frequency (to a few cm^{-1}) in different molecules. We refer to the latter type as a group frequency. This is the type we shall be using in this part of the unit since the presence of an absorption band at a particular frequency can be interpreted as the presence of a particular functional group.

It is normal after a little experience to treat the interpretation of an infrared spectrum in four regions, these are:

— 4000–2500 cm^{-1}

— 2500–2000 cm^{-1}

— 2000–1500 cm^{-1}

— 1500– 600 cm^{-1}

Let's take each of these regions in turn and discuss the most important and reliable absorption frequencies that could be present.

7.1.1.   The X—H Stretching Region (4000–2500 cm^{-1})

All fundamental vibrations in this region can be attributed to X—H stretching. Thus O—H stretch occurs at 3700–3600 cm^{-1}, if no hydrogen bonding is present and at lower frequencies in hydrogen bonded alcohols and acids, as you saw in Part Five.

N—H stretch also falls in this region, usually between 3400 and 3300 cm^{-1}. This absorption is usually much sharper than O—H stretch and can therefore be differentiated. Compounds containing the NH$_2$ group usually show doublet structure, while secondary amines show one sharp band, tertiary amines do not contain an N—H group and therefore do not absorb.

C—H stretching bands from aliphatic compounds occur in the range 3000–2850 cm^{-1}. They are moderately broad, but much sharper than hydrogen bonded O—H stretch. At least two peaks occur even with low resolution spectrometers. Most organic compounds have many C—H bonds resulting in medium intensity bands. Under high resolution it is often possible to analyse this region in more detail, eg the antisymmetrical and symmetrical C—H stretching absorptions of a methyl group usually occur at about 2965 and 2880 cm^{-1}, while the corresponding absorptions for a CH$_2$ group occur at 2930 and 2860 cm^{-1}.

Electronic effects of neighbouring groups can affect the frequency of these bands. The C—H stretching frequency of the aldehyde group, H—C=O, is often split into two bands at 2850 and 2750 cm^{-1}.

∏   We have discussed this splitting in Part Four. Can you recall why there are two bands?

This is an example of Fermi resonance between C—H stretch and the overtone of the C—H bending frequency.

If the C—H bond is adjacent to a double bond or aromatic ring the C—H stretching frequency increases and absorbs between 3100 and 3000 cm^{-1}.

This is very useful in distinguishing purely aliphatic compounds, but must be used with care, since in compounds containing a small number of 'aromatic' hydrogens and many aliphatic C—H bonds the peak(s) above 3000 cm^{-1} may only appear as a shoulder on the stronger aliphatic absorption and may be obscured. Hydrogens attached to carbons in small strained rings (cyclopropane and cyclobutane) and to carbons carrying chlorine also absorb above 3000 cm^{-1} and may also lead to confusion. However, evidence for the presence of an aromatic ring can be reinforced by a study of other parts of the spectrum as we shall see below.

Hydrogen attached to an sp hybridised carbon gives rise to higher frequencies and is easily interpreted, for example, C≡C—H stretch absorbs as a sharp, medium intensity, single peak at about 3300 cm^{-1}.

Deuterated compounds would be expected to show the C—D stretch at a factor of 1.41 less, ie at about 2130 cm^{-1}. These absorptions are usually a little higher than this, attributed to the failure of our model of totally independent vibrations.

SAQ 7.1a

Examine the spectra in Figs. 7.1a–7.1e and classify them as below:

(*i*)   aliphatic C—H bonds only,

(*ii*) aliphatic and aromatic C—H bonds,

(*iii*) an alkene or aromatic compound containing no aliphatic C—H bonds,

(*iv*) an alkyne,

(*v*)   a deuterated compound.

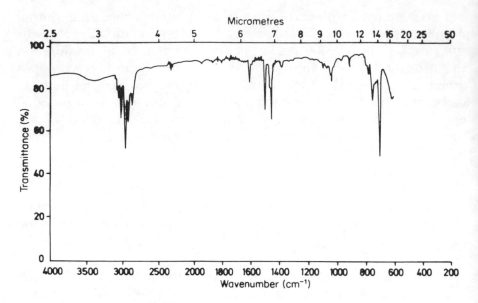

Fig. 7.1a. *Infrared spectrum of an unknown substance (SAQ 7.1a)*

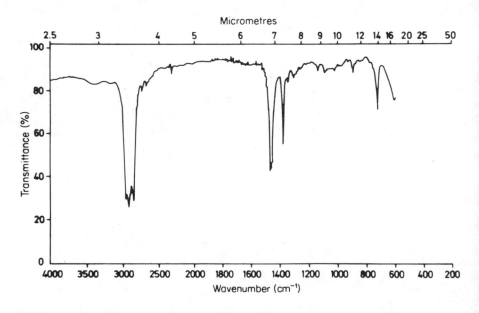

Fig. 7.1b. *Infrared spectrum of an unknown substance (SAQ 7.1a)*

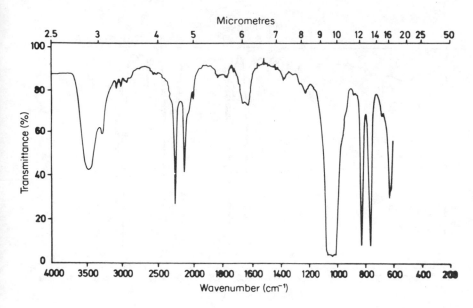

Fig. 7.1c. *Infrared spectrum of an unknown substance (SAQ 7.1a)*

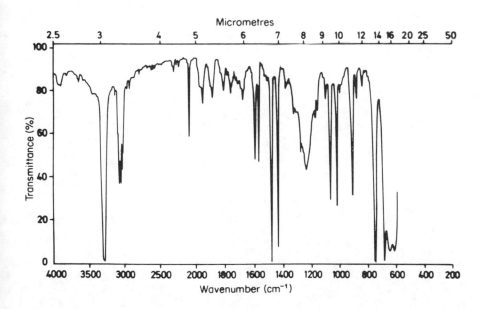

Fig. 7.1d. *Infrared spectrum of an unknown substance (SAQ 7.1a)*

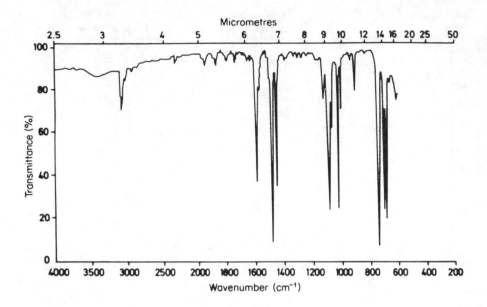

Fig. 7.1e. *Infrared spectrum of an unknown substance (SAQ 7.1a)*

7.1.2. The Triple Bond Region (2500–2000 cm^{-1})

Triple bond stretching absorptions fall in this region because of the high force constants of the bonds.

Carbon–carbon triple bonds absorb between 2300 and 2050 cm^{-1}, while the other common group in organic molecules, the nitrile group, C≡N, occurs between 2300 and 2200 cm^{-1}. These are usually easily distinguished since C≡C stretch is normally very weak, while C≡N stretch is of medium intensity.

∏   Can you recall why this should be so?

Well, the change in dipole moment during absorption in a C≡C triple bond is likely to be very small, unless a polar group is attached, while the C≡N group has a large dipole and hence a large change as the bond length is varied. The intensity of the absorption depends on this difference in dipole moment.

These are the only common absorptions in this region, but you may come across X—H stretching absorptions, where X is a more massive atom like sulphur, phosphorus or silicon. These usually occur at 2500, 2400 and 2200 cm^{-1}. The properties (and nasal attack) of these compounds means that confusion is unlikely however!

SAQ 7.1b

> Carbon monoxide absorbs at 2143 cm^{-1}. What does this tell us about the bond order in this molecule?

SAQ 7.1b

7.1.3. The Double Bond Region (2000–1500 cm^{-1})

The principal bands in this region are C=C and C=O stretch. We shall be studying the carbonyl group in some detail below, suffice to say here that the carbonyl stretch is one of the easiest absorptions to recognise in an infrared spectrum, is usually the most intense band in the spectrum and depending on the type of C=O, occurs in the region 1830–1650 cm^{-1}, but note that metal carbonyls may absorb above 2000 cm^{-1}.

C=C stretch is much weaker and often absent (for symmetry or dipole moment reasons) and occurs around 1650 cm^{-1}.

C=N stretch also occurs in this region and is usually stronger, but is much less common in organic molecules.

The N—H bending vibration in amines occurs between 1630 and 1500 cm^{-1} and is usually strong. Before assigning a band always check the N—H stretching region above 3000 cm^{-1} to avoid this possible confusion.

A series of weak bands in the region 2000–1650 cm^{-1} are often used for assignment purposes. These are combination bands from substituted benzenes. It has been found that the intensity, frequency and number of absorptions in this region are a reliable index to the substitution pattern in the benzene ring. I give typical patterns in Fig. 7.1f.

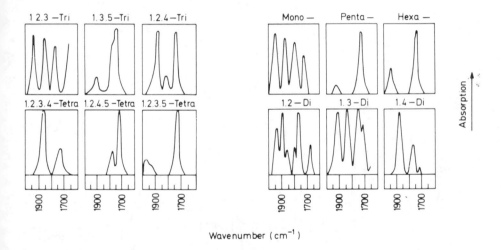

Fig. 7.1f. *CH Out-of-plane bending overtones and combination band patterns in 2000–1650 cm^{-1} region*

SAQ 7.1c

Examine the three spectra in Figs. 7.1g–7.1i.

They are of 1,2-dimethyl, 1,3-dimethyl and 1,4-dimethylbenzene, in the region 2000–1650 cm^{-1}.

Which is which?

Note that these are absorbance spectra.

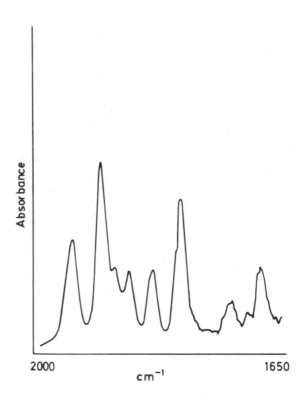

Fig. 7.1g.   *Infrared spectrum of a dimethyl substituted benzene*

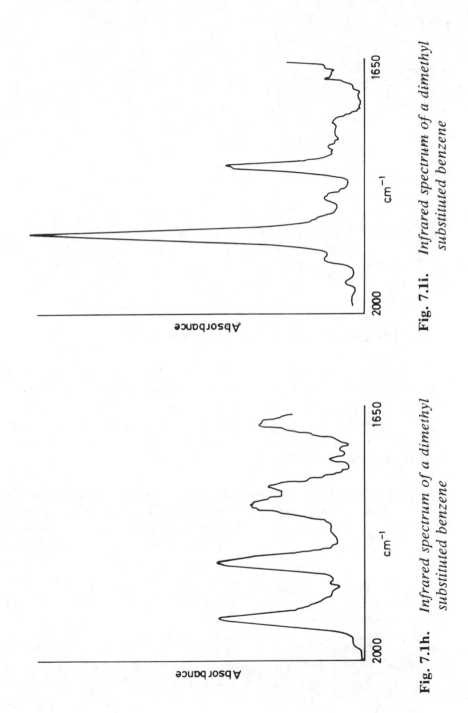

Fig. 7.1i.   *Infrared spectrum of a dimethyl substituted benzene*

Fig. 7.1h.   *Infrared spectrum of a dimethyl substituted benzene*

SAQ 7.1c

7.1.4.   The Fingerprint Region (1500–600 cm^{-1})

At frequencies corresponding to values greater than 1500 wavenumbers, it is usual to be able to assign each absorption band in an infrared spectrum. This is no longer true below 1500 cm^{-1}. This region is referred to as 'the fingerprint region' for this very reason, since very similar molecules give different absorption patterns at these frequencies.

Most single bonds absorb at similar frequencies and hence the vibrations couple. The observed pattern will depend on the carbon skeleton, and the resulting bands will originate from oscillation of large parts of the skeleton, or the skeleton and attached functional groups. Carbon–carbon stretching frequencies can also couple with C—H bending vibrations. It is perhaps surprising that any useful interpretive information can be got from this region. Many useful bands do exist however.

SAQ 7.1d   Would you expect C—O stretch to be more or less intense than C—C stretch?

SAQ 7.1d

The C—O stretching frequency is one of the bands that can be useful. If no intense band appears in this region, you can usually be sure that no C—O bonds are present. The frequency is rather variable, usually the absorption occurs between 1000 and 1400 cm^{-1}.

Aromatic rings give rise to two bands at 1600 and 1500 cm^{-1}. They are usually sharp, but are of variable intensity and occasionally the band at 1600 cm^{-1} splits into a doublet.

Aromatic rings and alkenes give rise to other bands which are perhaps the most useful in this region of the spectrum. These are out-of-plane C—H bending vibrations which occur between 1000 and 700 cm^{-1}.

In alkenes the pattern varies depending on the substitution pattern so that cis, trans and alkenes containing the $=CH_2$ group can be differentiated.

In substituted benzenes the spectral pattern gives information on the substitution pattern in the ring, since bands characteristic of one, two and three adjacent C—H bonds appear. This means that 1,2 substitution gives a different pattern from 1,3 and 1,4 substituted ring systems.

SAQ 7.1e   Absorption in the fingerprint region occurs at variable frequency for C—C and C—O stretching frequencies. Recalling the rules that were put forward in Part Four for coupling between vibrations, would you expect C—H out-of-plane deformations to couple with these stretching vibrations?

The nitro group, NO_2, gives two strong peaks at 1375 and 1550 cm^{-1}.

C—Cl stretch occurs around 700 cm^{-1} and is easily confused with C—H out-of-plane bending from aromatic rings.

The most important rule in spectral interpretation is to look at the whole spectrum, not spot a band and jump to conclusions. For example if the presence of a benzene ring is suspected, peaks should be present for C—H stretch above 3000 cm^{-1}, but the combination bands between 1600 and 2000 cm^{-1} and peaks for C—H bend between 600 and 1000 cm^{-1} should also be present. Checks should also be made for the peaks at 1600 and 1500 cm^{-1}.

There are a few general rules that can be stated to help you to use a vibrational spectrum in the determination of a structure; however, the most effective way to learn is through practice.

The guidelines are really common sense.

1. Look first at the high-frequency end of the spectrum (above 1500 cm^{-1}) and concentrate initially on the major bands.

2. For each band, short-list the possibilities using a correlation chart, see Appendix, and then use more detailed tables if necessary.

3. Use the low-frequency end of the spectrum for confirmation or elaboration of possible structural elements.

4. Do not expect to be able to assign every band in the spectrum.

5. Keep cross-checking wherever possible, eg an aldehyde should absorb near 1730 cm^{-1} *and* in the region 2700–2900 cm^{-1}.

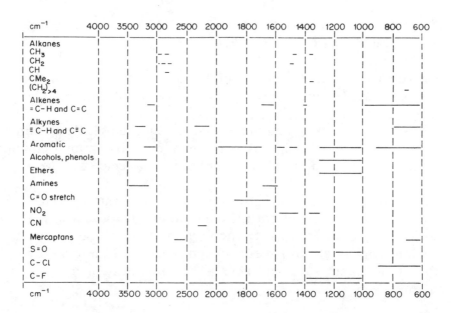

Fig. 7.1j.   *A simple correlation chart*

6. Place as much reliance on negative evidence as on positive evidence. Eg if there is no band in the 1600–1850 cm^{-1} region, it is most unlikely that a carbonyl group is present.

7. Band intensities should be treated with some caution; under certain circumstances they may vary considerably for the same group.

8. When looking for small frequency changes be wary; this can be influenced by whether the spectrum was run as a solid or liquid or in solution. If in solution, some bands are solvent sensitive.

9. Do not forget to subtract solvent bands or nujol bands. Do not confuse these with bands from the sample.

A simple correlation chart is given in Fig. 7.1j. I suggest you have a short look at it now, bearing in mind that the origin of the bands will be described in later sections of this unit. Certain bands will be highlighted and separate tables given later. All the correlation charts are collected together in an Appendix at the end of the Unit.

7.2.   A CLOSER LOOK AT ONE FUNCTIONAL GROUP – THE CARBONYL GROUP

I have tried to briefly survey useful ir bands in the section above. I have also tried to explain why they are useful in some cases.

I now want to look at one functional group in closer detail and try to explain the smaller shifts that are seen in its absorption frequency when other atoms in its vicinity within the molecule change.

I have chosen the carbonyl group (C=O) as an example, since it occurs in a wide variety of organic compounds and occurs with a wide variety of groups attached to the carbon atom. These groups have a wide range of electronic effects and makes this an ideal example to try to explain small but very important shifts in ir absorption frequencies.

First, let's make sure we understand the chemistry involved.

SAQ 7.2a   Which of the following compounds contain a carbonyl group? Look up their structures if you need to.

(*i*)   benzoic acid

(*ii*)   ethanal

(*iii*)   ethanol

(*iv*)   urea

(*v*)   pentan-2-one

(*vi*)   ethanoyl chloride

(*vii*)   acetic anhydride

(*viii*)   picric acid

(*ix*)   ethyl acetoacetate

The carbonyl group occurs in a very wide range of compounds, which chemists, over the years, have classified according to chemical properties and recognise as containing different functional groups. Esters behave differently from ketones which have a different chemistry from amides etc. It should not surprise you then, that the infrared absorption frequency of the carbonyl group in these compounds is also characteristic. The range is from 1860 cm^{-1} for acid fluorides to 1620 cm^{-1} for amides. This range is much larger than most other functional groups. The danger is that the range is overlapped by other functional groups, but, fortunately this is only true to a very limited extent. The $C{=}C$ stretching frequency (which is usually much weaker) and to a lesser extent $C{=}N$ stretch and N—H bending vibrations are the only ones. The carbonyl group has a region more or less to itself in the infrared, which is very fortunate for organic chemists.

Within any class of carbonyl compound the range is quite narrow. It is possible therefore, not only to recognise the presence of a carbonyl group in a compound, but to place it in one of the classes above.

I give a correlation chart below for the carbonyl region.

SAQ 7.2b

The following are examples of an ester, amide, ketone, aldehyde, acid chloride and acid anhydrides

CH_3COOEt CH_3CONH_2 CH_3COCH_3

CH_3CHO CH_3COCl $CH_3CO{-}O{-}COCH_3$

With reference to these examples and to the correlation table in Fig. 7.2a, place the functional group classes in order of increasing frequency for the $C{=}O$ stretch.

I will start you off: amide < $\longrightarrow$

SAQ 7.2b
(cont.)

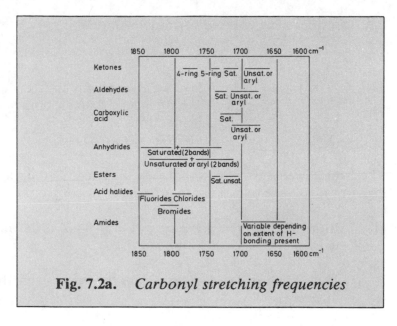

Fig. 7.2a. *Carbonyl stretching frequencies*

You should have noted in your examination of the correlation chart when answering SAQ 7.2b that there is a strong dependence of the stretching frequency on the group attached to the carbonyl carbon. Each class of compounds has a distinctive frequency.

Let's consider a generalised carbonyl compound;

R
　＼
　　＼
　　　C＝O
　　／
　／
X:

Where R is an alkyl group and X is a group
more electronegative than carbon carrying
lone pair electrons.
X could be N or O or Cl etc.

The group X can have effects that vary the length, strength and bond order of the C＝O bond. These effects are of course interrelated.

∏   If the bond length *decreases* will the bond order increase or decrease?

If the bond order *decreases*, will the bond strength increase or decrease?

We can say that as the bond order increases, the bond strength increases and the bond length decreases. Remember that the force constant increases as the bond strength increases, ie as the bond order increases. Any electronic effect that acts to shorten a bond will increase the frequency of vibration, the converse also being true.

SAQ 7.2c   Are the following statements true or false?

(*i*)   Double bonds are shorter than triple bonds and have larger force constants.

(*ii*)   Triple bonds absorb in the infrared region at shorter wavelengths than single bonds since they have larger force constants.

(*iii*)   the group X lowers the C—R bond order in X—C—R, it will raise its stretching frequency.

SAQ 7.2c

The effect of the group X can be explained for many carbonyl compounds using the well known electronic theories developed by organic chemists. These effects are usually referred to as inductive and mesomeric or resonance effects. Inductive effects act through the sigma electrons or bonds in a molecule, while the mesomeric or resonance effect acts through pi electrons or bonds. The inductive effect is usually given the symbol I with a sign indicating the direction of the effect, ie $+I$ is an electron donating group, while $-I$ is electron withdrawing. The symbols $+M$ and $-M$ are used in the same sense for the mesomeric effect.

In our generalised carbonyl compound the group X is more electronegative than carbon and therefore exerts a negative or $-I$ effect. Conversely the mesomeric or resonance effect is in the opposite direction, the lone pair electrons are donated *towards* the carbonyl group.

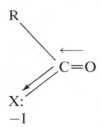

⊓   What influence will these effects have on the bond order, hence force constant, of the carbonyl group?

The negative inductive effect will pull the bonding electrons closer to the carbon atom, thus shortening the $C=O$ bond.

This will increase the bond order, increasing the absorption frequency.

The positive resonance effect which can be summarised using the resonance structures below:

R
>C=O ⟷ R
X: >C—O⁻
           X+

+M

or alternatively:

R
>C—O
δ⁺ X:

will lower the bond order thus decreasing the vibrational frequency.

SAQ 7.2d

You found in SAQ 7.2b that the carbonyl stretching frequency increased in the order amides, ketones, aldehydes, esters, acid chlorides. Use this information to deduce whether the inductive or resonance effect is the dominant effect

where X = O, N, Cl in the molecule

$$\begin{array}{c} R \\ \diagdown \\ \diagup \\ X \end{array} C{=}O$$

by comparison with the case where X = C (ketones).

These effects are additive and can be used to intuitively rank stretching frequencies. Try the following question.

SAQ 7.2e   Place the following compounds in order of *increasing* carbonyl stretching frequency.

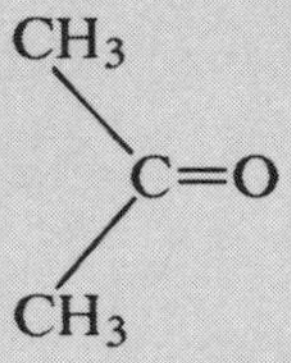

$$CH_3$$
$$\diagdown$$
$$C=O$$
$$\diagup$$
$$CH_3$$

(i)

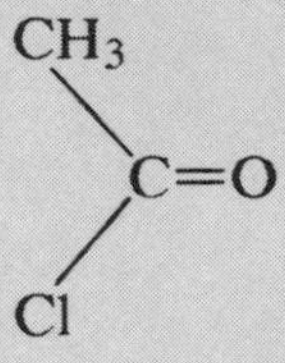

$$CH_3$$
$$\diagdown$$
$$C=O$$
$$\diagup$$
$$Cl$$

(ii)

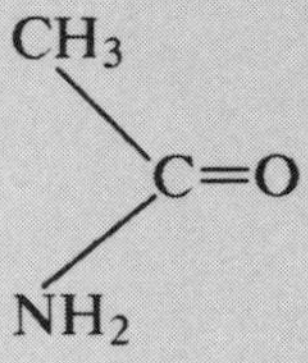

$$CH_3$$
$$\diagdown$$
$$C=O$$
$$\diagup$$
$$NH_2$$

(iii)

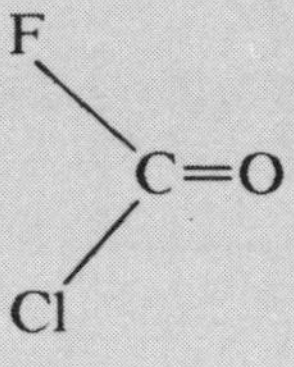

$$F$$
$$\diagdown$$
$$C=O$$
$$\diagup$$
$$Cl$$

(iv)

$$CH_3$$
$$\diagdown$$
$$C=O$$
$$\diagup$$
$$F$$

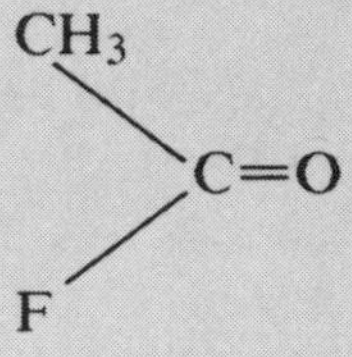

(v)

The resonance effect also dominates in conjugated compounds. So aromatic rings and double bonds conjugated to carbonyl groups lead to lower absorption frequencies by about 20–40 wavenumbers. The compounds below, methyl vinyl ketone (*a*) and acetophenone (*b*) absorb at a lower frequency than acetone.

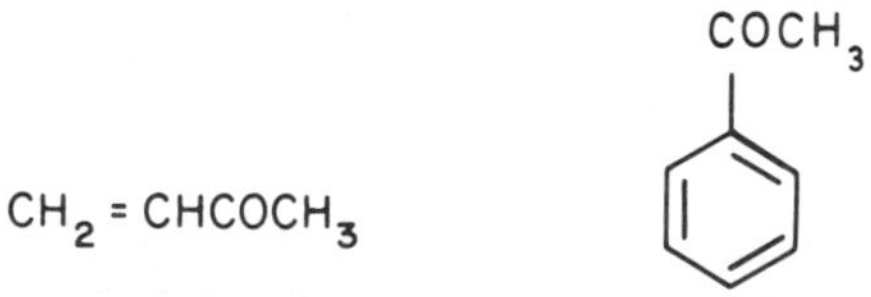
CH2 = CHCOCH3
COCH3

(a) (b)

These effects are all additive, so again educated guesses can be made as to absorption frequency for unknown compounds.

SAQ 7.2f

> Arrange the following molecules in order of increasing carbonyl stretching frequency,
>
> 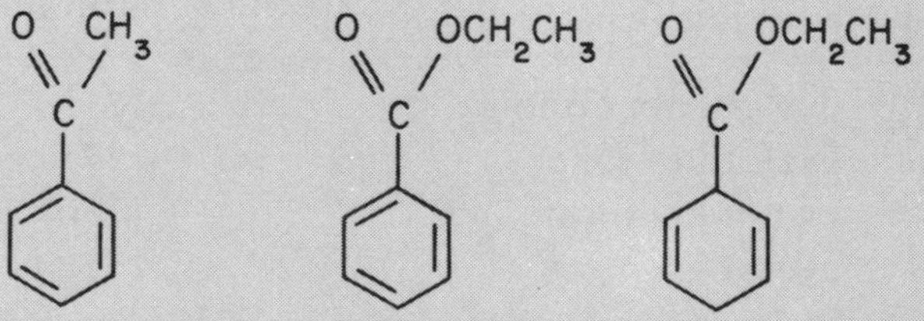
> O CH3
> C
> O OCH2CH3
> C
> O OCH2CH3
> C
>
> (a) (b) (c)

SAQ 7.2f

These simple ideas work very well for many compounds, but you must be careful, because there are other effects that can become important. We have assumed in the discussion above that the carbonyl group vibration is contained within the group and that no motion of any other part of the molecule is involved. This simplification breaks down occasionally.

Look at the carbonyl stretching frequencies of the four cyclic ketones in the table below.

Compound	Carbonyl frequency (cm^{-1})
cyclopropanone	1815
cyclobutanone	1775
cyclopentanone	1750
cyclohexanone	1718

In large rings the stretching frequency is the same as in open chain compounds, but as the ring size decreases the absorption occurs at higher wavenumber. We cannot invoke electronic effects here, nor are there vibrations in the same plane at similar frequencies so coupling in the sense that it was introduced earlier does not explain this either. We must find a new effect.

This is normally referred to as a mechanical effect and you should not confuse it with vibrational coupling. Let me try to explain.

In small rings (less than six carbon atoms) the ring is rigid. This is unlike open chain compounds and larger ring systems which are quite flexible, since neighbouring carbons can undergo small compensating displacements during the C=O stretch. Compression of the C=O in small ring systems results in compression of adjacent C—C bonds, more energy is required, and hence the stretching frequency increases as the ring becomes more rigid. This means that the C=O stretching frequency increases as the ring size decreases, as shown by the four cyclic ketones.

SAQ 7.2g Bearing in mind the typical carbonyl stretching frequencies of cyclic ketones for 3, 4, 5 and 6 membered ring systems (1815, 1775, 1750 and 1718 cm^{-1} respectively), can you make an educated guess at the C=O stretching frequency in the compound below?

SAQ 7.2g

SAQ 7.2h   Assign the appropriate C=O stretching frequencies to the structures below, from the following values:

1775,   1750,   1745,   1700   cm^{-1}

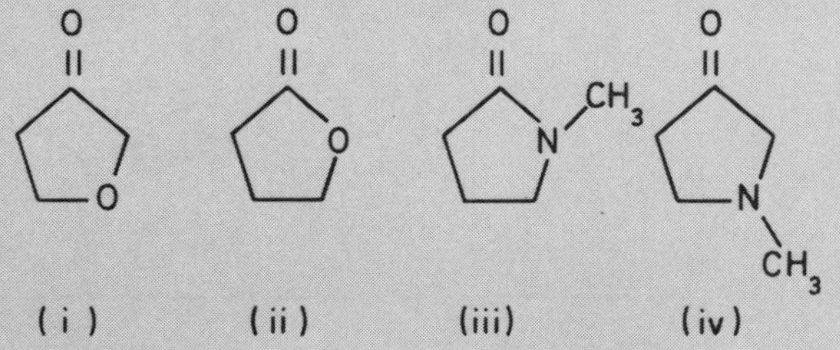

(i)   (ii)   (iii)   (iv)

SAQ 7.2h

These ideas can be applied to other groups besides carbonyl. Consider the nitrile stretching frequencies of ten compounds given in the table below.

Compound	Nitrile Stretch (cm^{-1})	Compound	Nitrile Stretch (cm^{-1})
CH_3CN	2267	CH_3CH_2CN	2265
FCN	2319	FCH_2CN	2266
ClCN	2215	$ClCH_2CN$	2256
BrCN	2200	$BrCH_2CN$	2249
ICN	2158	ICH_2CN	2248

The first thing to notice is that the five compounds on the left have large differences in absorption frequency, while those on the right show much smaller effects. This is exactly what you would expect, because the substituent that is being changed is bonded directly to the nitrile group in the left-hand side of the table, but is one carbon removed from the nitrile group in the right-hand side column.

Let us try and explain the large differences for the five compounds in the column on the left. If we assume that the differences can be explained solely by the inductive effect, we would not expect the order of increasing frequency to be the observed one: I, Br, Cl, CH_3, F.

∏   What would you expect?

Yes, CH_3 I, Br, Cl, F. The methyl group is electron donating $(+I)$, and this would lower the frequency. The halogens are all more electronegative than the methyl group and since F is of greatest electronegativity, the high nitrile stretching frequency for FCN is to be expected.

Obviously the inductive effect does not provide the mechanism for the actual order that is observed. Neither is it any use involving the resonance effect because the methyl group has no lone pair electrons. The only other effect we have come across is coupling. Unfortunately this does not provide a likely explanation, because the absorption frequencies of the $C\equiv N$ triple bond and the X—CN single bond (where X = C, I, Br, Cl and F) are very different and vary from group to group.

Let's consider what happens to the atom X when the triple bond vibrates. X can either move with the vibration, leaving the bond length X—C unchanged (case A below), or X can remain stationary, compressing the X—C bond as $C\equiv N$ lengthens, stretching X—C as $C\equiv N$ compresses (case B).

    ⟵   ⟵   ⟶   ⟵   ⟶

X ——— C ——— N   X ——— C ——— N

     Case A   Case B

As the atom X becomes more massive inertia effects will make case B more and more likely. Does this explain the frequencies above? How could we test this hypothesis?

If the hypothesis is correct then the vibrational frequency of $C\equiv N$ should correlate with the 'stiffness' of the X—C bond. We have a measure of this 'stiffness' in the force constant of the X—C bond. It has been shown that a plot of force constant against absorption frequency is linear. This proves our hypothesis. This effect is referred to as *adjacent-bond interaction* and can be seen in many other molecules, for example alkenes. It turns out that only the least massive element, hydrogen and its isotopes, move with the vibration of adjacent bonds.

You are by now getting expert at assigning $C=O$ stretching frequencies and explaining mechanical and electronic effects in this class of compound. I have dealt with the $C=O$ group in a lot of detail since you will find it so very useful and the knowledge can be transferred to other groups. You will be meeting it again in the next section when I deal with compounds by chemical class. Before you leave this section, it might be a good idea to have a good long look at the carbonyl correlation table (Fig. 7.2a) and try to rationalise the small shifts observed for this group.

7.3. CHARACTERISTIC ABSORPTIONS OF THE COMMON CLASSES OF ORGANIC COMPOUNDS

I hope this section will enable you to bring together information from other Parts of the Unit and apply it to the infrared spectra of particular classes of organic compound. You may therefore need information from earlier parts of the Unit to answer some of the self assessment questions. It should also provide a good reference source for information at a future date. In other words, don't try to remember all the details at this stage; much of it is only there because I have tried not to leave unanswered some of the questions that may occur to you. You will also find a summary of all the information on absorption frequencies at the end of the section.

7.3.1. Alkanes

These compounds contain only C—H and C—C bonds and their spectra therefore contain few bands.

Considering the two types of stretching vibrations C—H and C—C, which of these occurs at highest frequency and which is the most useful?

The C—C stretch occurs in the skeletal region, is usually very weak, couples with other C—C stretching vibrations and is therefore of little use diagnostically.

C—H stretch is a useful absorption, occurring in saturated hydrocarbons below 3000 cm^{-1} and is usually of medium intensity.

Three sorts of bending modes are possible, C—C—C, C—C—H and H—C—H. Only the last of these is useful. The C—C—C bending mode absorbs outside the normal range, and the C—C—H can be useful but often occurs in the skeletal region. This leaves the most useful absorption, H—C—H. This occurs in the regions 1320–1480 cm^{-1} and 700–1200 cm^{-1}. The former are deformation bands and the latter rocks, wags and twists. Note the large range of the latter, making this less than ideal for interpretation. This is again due to coupling with other vibrations.

Let's first look at the stretching vibrations. In saturated hydrocarbons the hydrogens can only be part of a methyl (CH$_3$), methylene (CH$_2$) or methine (CH) grouping. Two stretching frequencies are possible for CH$_3$ and CH$_2$ groups, can you recall what they are?

They are symmetrical and anti-symmetrical stretching modes

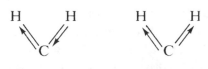

Anti-symmetric   Symmetric
stretch   stretch

The stretching frequencies of the CH_2 group are generally the same in open chain and saturated cyclic compounds. They occur at 2926 and 2853 cm^{-1}. Methyl groups absorb at 2962 and 2872 cm^{-1}. They seldom vary by more than 10 cm^{-1}.

The resolution of many instruments is not good enough to resolve these four bands into more than a doublet. Modern grating spectrometers can resolve them if the spectrum is recorded with a narrow slit width.

∏　　How many stretching bands would you expect from a methine group?

Only one is possible, this is generally weak and occurs around 2890 cm^{-1}.

You may recall from Part Four that bending modes are a little more complex. They are classified into deformations, rocks, wags and twists. Deformations are the most useful of these since the others either occur outside the normal range, or couple with other parts of the molecule and are therefore of very variable frequency. Methylene groups have only one possible deformation, while methyl groups have two.

The symmetrical CH_3 deformation is usually at 1380 cm^{-1}, the antisymmetrical one at about 1450 cm^{-1} and the CH_2 deformation at 1465 cm^{-1}.

SAQ 7.3a	Examine the spectrum of nonane in Fig. 7.3a and describe the vibrations corresponding to the absorptions marked A, B and C. ⟶

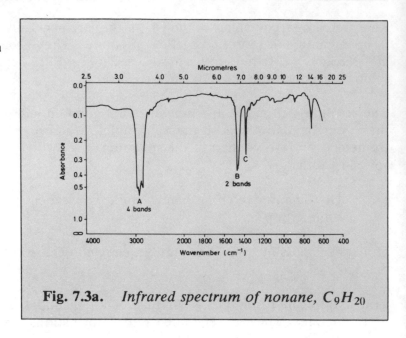

Fig. 7.3a. *Infrared spectrum of nonane, C_9H_{20}*

Rocking and wagging vibrations are usually of little use for interpretation purposes, but one exception is a band at 720 cm^{-1} which occurs in compounds containing the $(CH_2)_n$ group where n is greater than four.

One last useful correlation is that ascribed to the gem-dimethyl group, $C(CH_3)_2$. Any compound containing this group has a doublet at about 1375 cm^{-1}.

SAQ 7.3b

From the list of compounds below, choose those that

(*i*)   absorb at 720 cm^{-1},

(*ii*) absorb at 1375 cm^{-1},

(*iii*) absorb at both 720 and 1375 cm^{-1},

(*iv*) do not absorb at either 720 or 1375 cm^{-1}.

2-methylbutane

cyclopentane

2-methyloctane

3-methylpentane

butane

1,1-dimethylcyclohexane

SAQ 7.3b

This section is more important than you may at first think, since many compounds contain saturated alkyl chains. This information is equally applicable to them. Even if they contain functional groups, only the CH_2 group adjacent to the functional group is affected. It is usually easy to spot gem-dimethyl groups, chains of CH_2 groups and alkyl C—H stretching.

7.3.2. Alkenes

These compounds contain the C═C group, the majority having hydrogen attached to the double bond. Four vibrational modes are associated with this molecular fragment. These are out-of-plane and in-plane C—H deformations, C═C stretch and ═C—H stretch.

Refer back to Part Four for a detailed study of the C═C stretching frequency and factors affecting its intensity. You will also find a treatment of ═C—H deformation bands.

The number of ir active modes is symmetry-dependent and a plane of symmetry through the double bond, for example, makes the C=C stretch inactive. No matter what the symmetry, however, at least one of the four bands *will* be active.

Let's first look at C=C stretch. This mode absorbs in the range 1600–1680 cm^{-1} and is usually weak. Some useful generalisations can be made.

(*a*) We saw above that the absorption frequency of the carbonyl group was decreased by 20–40 cm^{-1} by conjugation, a similar effect operates here. We also observe an increase in the intensity of the absorption band. This conjugation effect also applies when atoms with lone-pair electrons are bonded directly to the alkene. Hence the C=C stretching frequency is lowered when a halogen is directly bonded, except for fluorine when the opposite is observed (and the frequency is increased).

(*b*) Cyclohexene absorbs at 1646 cm^{-1}, which is identical to open chain alkenes. As the ring size decreases the frequency also decreases (to 1611 cm^{-1} in cyclopentene and 1566 cm^{-1} in cyclobutene). Exocyclic double bonds increase in frequency as ring size decreases, exactly as in cyclic ketones and for the same reason. Think about it! (and refer back if you feel the need).

SAQ 7.3c   Arrange each of the two sets of compounds below in order of increasing C=C stretching frequency.

(*i*)

$$\underset{\text{(a)}}{\diagup\!\!\!\overset{\diagdown}{C}=CH_2} \quad \underset{\text{(b)}}{\diagup\!\!\!\overset{\diagdown}{C}=CBr_2} \quad \underset{\text{(c)}}{\diagup\!\!\!\overset{\diagdown}{C}=CHF}$$

$\longrightarrow$

SAQ 7.3c
(cont.)

(*ii*)

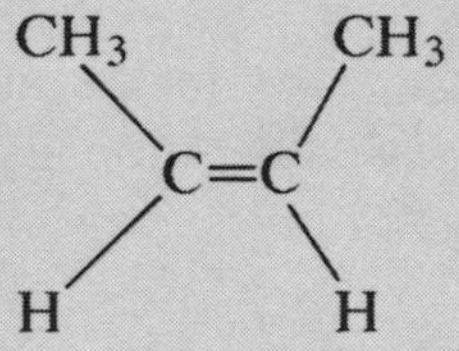

CH$_3$ CH$_3$

C=C   cyclopentene cyclobutene

H   H

(a)   (b)   (c)

In-plane =C—H bending absorptions occur near 1400 cm^{-1} and the out-of-plane modes absorb between 600 and 1000 cm^{-1}. The latter are very helpful in assigning unsaturation to a molecular structure, but the former are usually of little help, since they tend to be in a 'busy' region of the spectrum. Many other vibrations are active at this frequency (1400 cm^{-1}).

The out-of-plane bending vibrations are of even more help, since they can often give us extra structural information. Study Fig. 7.3b below, where the characteristic absorption frequencies of alkenes of different types are detailed.

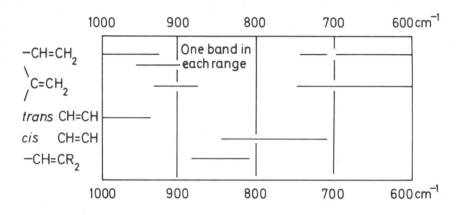

Fig. 7.3b. *Out-of-plane bending vibrations of alkenes*

Note Halogen substitution may lower the absorption frequency sufficiently to take it outside the ranges quotes above.

SAQ 7.3d Which of the pairs of alkenes given below could you distinguish using the spectral information provided by out-of-plane bending vibrations of alkenes?

(*i*) *cis*-But-2-ene and But-1-ene,

(*ii*) *trans*-Hex-2-ene and Cyclohexene

(*iii*) 2,3-Dimethyl-but-2-ene and 2-Methyl-but-2-ene

(*iv*) Propene and 2-Methylpropene

SAQ 7.3d

Lastly the C—H stretching frequency in unsaturated compounds occurs at higher frequencies than in saturated molecules. It should become second nature to check the C—H region around 3000 cm^{-1}. If unsaturation is present absorption will occur above 3000 cm^{-1}. The normal range for =C—H stretch in alkenes is 3000–3100 cm^{-1}. Compare the two spectra in Fig. 7.3c. One is the C—H stretching region of cyclohexane, the other is of cyclohexene. It is immediately obvious which is which.

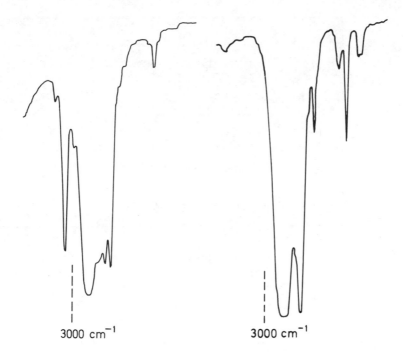

Fig. 7.3c. *The infrared spectra of cyclohexane and cyclohexene*
(C—H str region)

7.3.3. Alkynes

Three characteristic absorptions can be present, C—H stretch, C—H bend and C≡C stretch, depending on the structure of the compound.

The C≡C stretch occurs between 2100–2300 cm^{-1} and is usually very weak.

SAQ 7.3e   Which of the following molecules would be ex-
            pected to have the strongest C≡C stretch?

            (*i*)   Ph—C≡C—H

            (*ii*) CH_3—C≡C—CH_3

            (*iii*) CH_3CH_2—C≡C—CH_3

            Which would be weakest?

The stretching frequency of a hydrogen attached to a triple bonded
carbon occurs at very high frequency (3220–3320 cm^{-1}), is sharp,
can be fairly intense and is usually easy to assign. There is a possibility of overlap by O—H and N—H stretch in hydrogen bonded
compounds, but this can be overcome by recording the spectrum at
high sample dilution.

The C—H bending frequency is between 600 and 700 cm^{-1}. Fig. 7.3d is the spectrum of phenylethyne, can you find the three bands (C≡C str, C—H str and C—H bend) referred to above? I hope so!

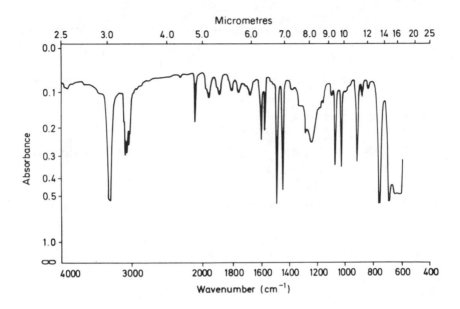

Fig. 7.3d. *Infrared spectrum of phenylethyne*

7.3.4. Benzenoid Compounds

The presence of a benzene ring in a molecule gives rise to absorption in six regions of the spectrum.

(*a*) 3000–3100 cm^{-1} C—H stretch

There is the possibility of confusion with alkene C—H stretch, but this can usually be sorted out by the presence or absence of other bands.

(*b*) 1650–2000 cm^{-1} overtone and combination bands (weak).

These are weak bands and need a liquid film spectrum to be reliable. Alkenes may also absorb in this region, but their bands are very much weaker. Some ir spectroscopists use the bands in this region as a reliable indication of substitution pattern in the ring, as we saw in Section 7.1.3. I like to look at both this region and that covered in (*f*) below.

(*c*) 1550–1600 cm^{-1} ring stretching (very variable intensity) sometimes two bands.

A peak is usually present at 1600 cm^{-1} but is of very variable intensity, depending on the polarity of the groups in the ring. A second peak may be present.

(*d*) 1450–1500 cm^{-1} ring stretching (very variable intensity) sometimes two bands.

The presence of this band, or bands (there may be two), is perhaps

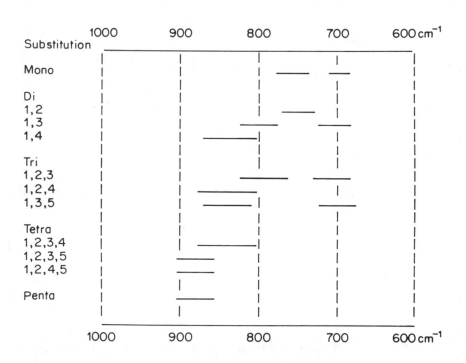

Fig. 7.3e.   *Out-of-plane bending vibrations of benzenoid compounds*

the most reliable indication of the presence of a benzene ring.

(*e*) 1000–1300 cm^{-1} C—H in plane bending (usually weak).

These peaks are in the fingerprint region, are of very variable frequency and tend to be very weak. They are not reliable bands for structure determination.

(*f*) 600–900 cm^{-1} C—H out-of-plane bend (usually strong).

These bands are strong and characteristic of the number of hydrogens in the ring, and hence can be used to give the substitution pattern. This information is summarised in Fig. 7.3e above.

It is best not to rely on these bands too much for structure determination, so always try to get other evidence such as that provided by ^{1}H nmr spectroscopy.

SAQ 7.3f   Examine the three spectra in Figs. 7.3f–7.3h. They are isomeric disubstituted benzenes. Which is 1,2 which 1,3 and which 1,4?

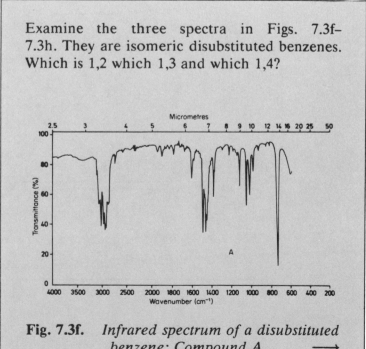

Fig. 7.3f.   *Infrared spectrum of a disubstituted benzene: Compound A* $\longrightarrow$

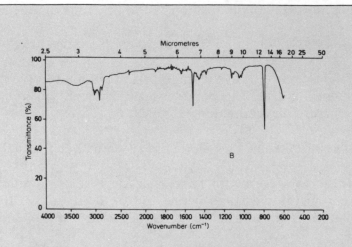

Fig. 7.3g.   *Infrared spectrum of a disubstituted
                  benzene: Compound B*

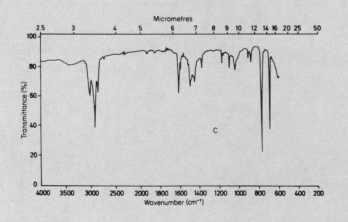

Fig. 7.3h.   *Infrared spectrum of a disubstituted
                  benzene: Compound C*

SAQ 7.3f

7.3.5. Alcohols, Phenols and Ethers

All three types of compound in the heading above contain the C—O stretching vibration. This occurs in the fingerprint region, couples with other modes and is of variable frequency, usually absorbing between 1000–1300 cm^{-1}. Its redeeming feature is that it is usually the most intense band in the spectrum and can be used, therefore, for compound identification.

Alcohols also contain the O—H stretch and the C—O—H bend. The former has been dealt with in detail in Part Five, and suffice to say here that it can be broad or weak depending on sample concentration. It absorbs between 2500 and 3650 cm^{-1} depending on molecular association and can give valuable structural information.

The C—O—H bending vibration occurs in the fingerprint region and is of little use to us.

SAQ 7.3g	Given that a molecule contains oxygen, how would you use an infrared spectrum to find out whether the compound was an acid, phenol, alcohol or ether?

You are now about half way through this Part. It may be a good time for a break! If you have understood the responses to the self assessment questions you are well on the way to becoming expert in spectrum interpretation.

7.3.6. Amines

In addition to hydrocarbon frequencies, amines would be expected to give N—H stretch and N—H bending absorptions.

Amines are classified as primary, containing the NH_2 group, secondary, containing the NH group and tertiary containing no hydrogen attached to nitrogen. This can alternatively be looked at via the number of C—N bonds. Here primary amines contain one C—N bond, secondary, two and tertiary, three.

Typical compounds are given below;

$$CH_3-NH_2$$

$$\overset{\displaystyle CH_3}{\underset{\displaystyle CH_3}{\diagdown}}N-H$$

$$\overset{\displaystyle CH_3}{\underset{\displaystyle CH_3}{\diagdown}}N-CH_3$$

Primary   Secondary   Tertiary

SAQ 7.3h

Do you consider it possible to differentiate the three classes of amine (primary, secondary and tertiary) using N—H stretching absorption?

Two absorptions occur in primary and secondary amines, the N—
H stretching and N—H bending modes. As you saw in SAQ 7.3h
this can be used to classify amines. The N—H stretching absorption
occurs between 3180 and 3490 cm^{-1}. It is usually sharp and is one
line in secondary amines and two lines (sometimes three) with a
spacing of 100 cm^{-1} in primary amines.

It should be noted that this is in the O—H stretching region. Little
confusion occurs since:

— the N—H stretch is usually sharp,

— dilution has little influence on position or sharpness.

This latter effect is because of the much weaker hydrogen bonding
present in amines compared to alcohols.

The N—H bending absorption occurs between 1580 and 1650 cm^{-1}
and is of medium intensity. It is usually stronger and broader than
other absorptions in this region. Can you recall any other bands
likely to occur here?

The C=C str occurs here from alkenes and ring stretching absorptions from benzenoid compounds.

7.3.7.   Aldehydes and Ketones

These compounds contain the carbonyl group which we considered
in detail in Section 7.2 above. There we concerned ourselves with
changes induced in the absorption frequency by the electronic and
mechanical effects of adjacent groups. Here and in later parts of
Section 7.3, we shall look at the molecule as a whole and show
how other frequencies can be used to identify carbonyl containing
compounds.

SAQ 7.3i

> Look back at the correlation chart for the C=O stretching frequency, Fig. 7.2a. What other classes of compound absorb in the same range as saturated aldehydes and ketones? Is it possible to distinguish these compounds using other absorptions likely to be present in the spectrum?

Π   The 'base' frequency for a saturated ketone is 1715 cm^{-1}. Can you recall two structural features that cause shifts from this value?

The presence of conjugation leads to a lower frequency of absorption, while the

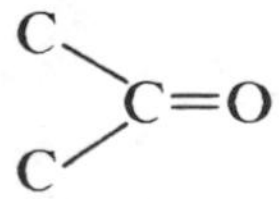

C\
   \C=O   angle also has an influence. You should recall that
C/

six-membered ring compounds absorb at the 'base' frequency, while smaller ring compounds absorb at higher values.

Another effect you may have recalled from Part Five is that hydrogen bonding can also exert an influence on this stretching frequency. This can easily be seen in hydroxy- and amino-ketones, particularly if the groups are in the correct stereochemical relationship for intramolecular hydrogen bonding.

These effects apply in exactly the same manner to aldehydes.

SAQ 7.3j

> The 'base' frequency for an open-chain or saturated ketone is 1719 cm^{-1}. The frequency of an aldehyde carbonyl group is found to be higher than this. Explain in a sentence why this is so.

The carbonyl stretching frequency regions of these two classes of compound overlap, however, when conjugative effects are taken into account. The C—H stretching frequency present in aldehydes comes to our rescue. You may recall that this frequency is very different from all other C—H stretching absorptions. It occurs between 2700 and 2900 cm^{-1} and is usually a doublet because of Fermi resonance with bands around 1400 cm^{-1}.

SAQ 7.3k

Examine the spectrum of 2-methoxybenzaldehyde in Fig. 7.3i. Can you see that the band for the aldehyde C—H str is a doublet, even though there is overlap of the high frequency peak with the alkane C—H str? I hope so.

Describe the effect responsible for the doublet formation.

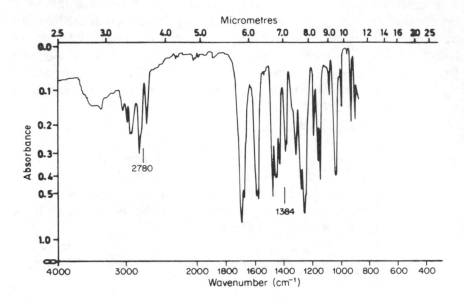

Fig.7.3i.   *The infrared spectrum of 2-methoxybenzaldehyde*

7.3.8.   Carboxylic Acids

We have looked at carboxylic acids in Part Five in the context of hydrogen bonding, hence you are already aware that this effect dominates the C=O stretching frequency in these compounds. It is found, however, that the base frequency for C=O stretch in monomeric aliphatic unsaturated acids occurs at 1760 cm^{-1}. You saw above that the base frequency in ketones was 1715 cm^{-1}, carboxylic acids therefore show a shift to higher frequency. How do we explain this?

Two electronic effects are present:

a negative
inductive effect   resonance effect

The inductive effect pulls electrons from the $C{=}O$ group, increasing the bond order and hence increasing the absorption frequency. The resonance effect reduces the bond order, leading to the opposite effect. In acids the inductive effect must be dominant.

If spectra of acids are obtained under normal conditions, either as a mull in the solid state or in solution at normal concentration, the $C{=}O$ stretching frequency is found around 1710 cm^{-1}. This can be explained by the presence of a hydrogen bonded dimer.

SAQ 7.31   Monomeric carboxylic acids show $C{=}O$ stretch at 1760 cm^{-1}, while in the solid state this absorption occurs at 1710 cm^{-1}. Can you explain this in terms of hydrogen bonding?

Conjugation of adjacent unsaturation leads to lower $C=O$ frequencies, but to a lesser extent than in aldehydes and ketones, the range is normally 1680–1715 cm^{-1}.

Hydrogen bonding also leads to broadening of the $O-H$ stretching absorption. This occurs over a very wide range, 2500–3500 cm^{-1}.

Other bands that can usually be assigned are $C-O$ stretch around 1200–1300 cm^{-1}, $C-O-H$ in-plane around 1400 cm^{-1} and $C-O-H$ out-of-plane bend at 800–1000 cm^{-1}.

The first two couple to $C-C$ stretching frequencies in the same plane and are therefore variable. The last of the three absorptions does not couple and is therefore useful in structure determinations. Note that the bending vibrations are analogous to similar absorptions in alkenes.

Carboxylic acids form salts. The infrared absorptions in these salts would be expected to be very different to the parent acids. The $O-H$ stretching frequency will be absent and the $C=O$ stretching frequency will also be modified. Try the question below before we go on to the next section.

SAQ 7.3m

The carboxylate anion can be written,

The $C-O$ bonds are therefore equivalent and have the same bond order, intermediate between double and single bonds. Which of the following would you expect to be correct for the $C-O$ absorption? $\longrightarrow$

SAQ 7.3m
(cont.)

> (*i*) Very intense single peak at 1500 cm^{-1}.
>
> (*ii*) Very intense single peak at 1800 cm^{-1}.
>
> (*iii*) Two bands, one at 1600 and one around 1400 cm^{-1}.

7.3.9. Esters and Lactones

The two most polar bonds in these molecules are the $C=O$ and $C-O$ bonds, and we would therefore expect strong absorptions from them. This is the case and these are the strongest peaks in the spectrum of any ester or lactone. Saturated esters absorb between 1735 and 1750 cm^{-1} for the $C=O$ group; conjugation lowers this by about 20 cm^{-1}. You should note that the $C=O$ stretching frequencies of esters, carboxylic acids and ketones overlap, how would you distinguish them?

Carboxylic acids have a broad O—H absorption, so can easily be distinguished. Esters and ketones can be told apart because of the strong C—O band in esters – this is variable but is usually very strong. Ketones also absorb in the fingerprint region but never with high intensity.

Lactones are simply cyclic esters. It is found, however, that the C=O stretching frequencies of lactones increase (unlike esters) if conjugation is present next to the singly bonded oxygen. It is thought that this is because the oxygen lone pair is in conjugation with both the carbonyl group and the other multiple bond. Loss of full conjugation to the carbonyl group therefore leads to an increase in carbonyl frequency. This cannot be the whole explanation however.

SAQ 7.3n

Arrange the following lactones in order of decreasing carbonyl frequency.

7.3.10.   Acid Anhydrides, Acid Halides and Amides

Acid anhydrides are usually easily distinguished from other car-bonyl containing compounds, because the C=O stretching fre-quency is invariably a doublet. Why do you think this is?

I suppose it could have been due to Fermi resonance but is in fact a result of symmetrical and anti-symmetrical stretching modes:

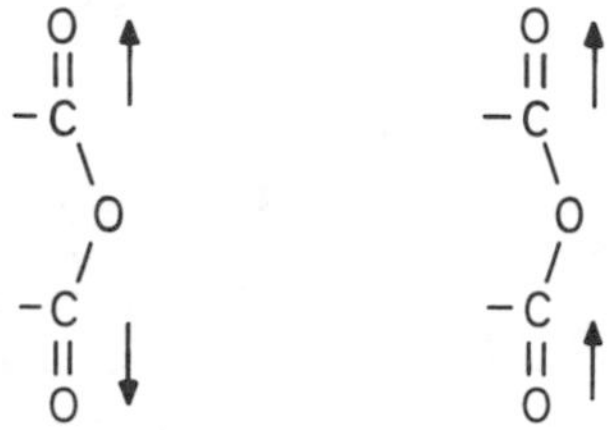

The same general rules apply here as above, regarding conjugation and ring size etc. Try the following.

SAQ 7.3o

Look at the structures below and assign each of the following pairs of carbonyl absorptions to one of the compounds, 1865, 1780; 1815, 1750; 1775, 1720 cm^{-1}.

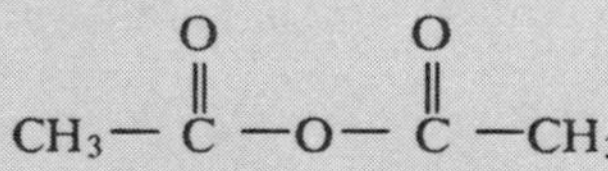

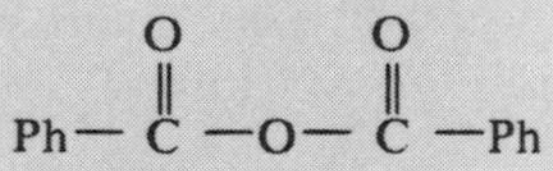

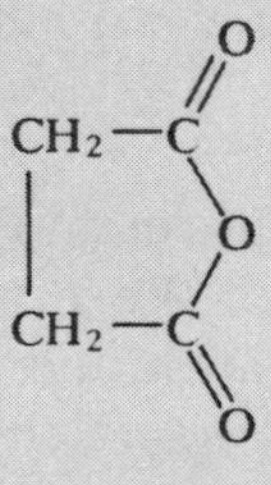

SAQ 7.3o

Acid anhydrides also absorb strongly in the fingerprint region due to the $C-O$ absorption.

Acid halides were covered in Section 7.2 and of course are easily distinguished because of their characteristically high absorption frequency. Can you recall why the $C{=}O$ stretching frequency is high in these compounds?

It is because of the dominant inductive effect and we would therefore expect acid fluorides to absorb at even higher frequency than acid chlorides.

In amides the carbonyl stretching frequency is lower than in ketones. The inductive effect of the nitrogen must therefore have a lesser effect than the resonance effect of the nitrogen lone pair. This may be because the electronegativity difference between nitrogen and carbon is less than for carbon and oxygen in acid anhydrides, or carbon and a halogen in acid halides.

The carbonyl stretching frequency is also lowered in the solid state by hydrogen bonding as in acids, but to a lesser extent. It is also affected by conjugation as in the other classes of compounds examined above.

The N—H stretching frequency can be used to classify amides as primary, secondary or tertiary just as we did for amines in Section 7.3.6. Care needs to be exercised here, however, since hydrogen bonding may broaden this absorption and smooth a doublet to a singlet. Spectra, however, must be recorded in dilute solution to be unambiguous.

The N—H in-plane bending frequency in primary and secondary amides occurs between 1510 and 1650 cm^{-1}.

SAQ 7.3p

Examine the structures below and choose one of the following carbonyl stretching frequencies for each, 1650, 1660, 1685 cm^{-1}.

The samples were at high dilution in a non-polar solvent.

(*i*) $CH_3.NH.CO.Ph$

(*ii*) $Ph.NH.CO.CH_3$

(*iii*) $CH_3.NH.CO.CH_3$

SAQ 7.3p

Before we leave carbonyl compounds try the following revision self assessment question.

SAQ 7.3q

Match the following compounds with the spectral details below.

Ph.CH$_2$.CO.CH$_3$

4-Methylbenzaldehyde

Ph.CH=CH.CO$_2$H

Ph.CH$_2$.CO.NH$_2$

(*i*)   2800, 2680, 1680, 750, 810, no absorption above 3100 cm^{-1}

(*ii*)   3500–2500 (broad), 1680, 1630, 710, 760 cm^{-1}

(*iii*) 3400, 3200, 1640, 690, 740 cm^{-1}

(*iv*)   1710, 680, 740, no absorption above 3100 cm^{-1}

SAQ 7.3q

7.3.11. Other Nitrogen-containing Compounds

We shall confine ourselves to only a few major classes.

The nitrile or cyanide group, $C\equiv N$, absorbs between 2240–2270 cm^{-1} in alkyl compounds, and as you should by now suspect, this is lowered by conjugation. It is more intense than the $C\equiv C$ stretching frequency and can usually be easily distinguished.

A very common group in aromatic compounds is the nitro group, NO_2, this is best written as a resonance hybrid;

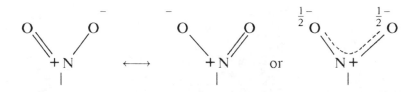

Π How many bands would you expect from the $N-O$ stretching vibration of this group?

It is exactly analogous to the CH_2 or NH_2 groups, and I would therefore expect a doublet. This is the case and the bands are found at about 1550 and 1370 cm^{-1} in saturated aliphatic nitro compounds. The bands are usually very intense because of the large dipole moment present. Conjugation lowers these frequencies to an extent which is determined by the electron donating or attracting power of the attached group.

SAQ 7.3r

Match the following compounds with the N—O stretching frequencies given.

Nitrobenzene

4-Nitroaniline

4-Nitrobenzaldehyde

(*i*)   1480 and 1319 cm^{-1}

(*ii*) 1560 and 1360 cm^{-1}

(*iii*) 1520 and 1355 cm^{-1}

Hydrazones and oximes contain the C=N bond, whose stretching frequency occurs over a wide range, as does the carbonyl group. In fact it occurs anywhere between 1400 and 1700 cm^{-1}. The N=N stretching frequency in azo-compounds is less variable and occurs around 1570 cm^{-1}.

7.3.12. Sulphur Containing Compounds

The S—H stretching frequency is very weak and may even be too weak to observe. It occurs at lower frequency than O—H, N—H or C—H, usually between 2520 and 2600 cm^{-1}. Any chemist can therefore usually detect the S—H group more easily with his nose than with a sophisticated spectrometer.

The oxygen-containing sulphur compounds are more complex. The S=O stretching in sulphoxides (R.SO.R) gives an intense band between 1000 and 1100 cm^{-1}.

Sulphones (R.SO$_2$.R) also contain the S=O group. How can we distinguish the two classes of compound?

The sulphone group will have two absorption frequencies analogous to the nitro group corresponding to symmetric and antisymmetric stretching. The actual frequencies observed are 1120–1160 and 1300–1350 cm^{-1}. These frequencies are also observed in sulphonic acids which also show O—H stretching absorption. Sulphonamides contain the SO$_2$NH$_2$ group and absorb in the same range. The opposing inductive and resonance effects of the nitrogen atom must therefore balance.

In sulphonyl chlorides a similar effect is seen as in the carbon acids, ie a move to higher frequencies for the S=O stretches. These compounds absorb about 30–50 cm^{-1} higher than sulphones or sulphonic acids.

7.3.13. Halogen Containing Compounds

The C—Halogen stretching absorption is usually intense, but occurs at low frequency, often outside the normal range. C—Cl can easily be confused with C—H bending frequencies from aromatic rings. The more massive the halogen, the lower the frequency. The self assessment question below illustrates these points.

SAQ 7.3s

The C—Cl stretching frequency occurs in the range, 600–850 cm^{-1}. Which of the following statements are true?

(i)   C—Br stretch occurs between 1000 and 1100 cm^{-1}.

(ii) C—I stretch occurs below the normal frequency range.

(iii) C—F stretch occurs in the range 1000–1400 cm^{-1}.

(iv) Absorptions due to C—F stretch are more intense than those due to C—I stretch.

I have collected the main absorption frequencies which are useful for structure determination below. This summary should prove useful to you and should be referred to until you have enough experience of spectrum interpretation to have memorised the main frequencies.

The next section gives you a lot of practice and you will not only need the summary but correlation tables also.

The correlation tables are collected together in the Appendix.

7.3.14. Summary of Characteristic Absorptions

ALKANES

(*a*) C—H stretch

CH_3, two bands at 2962 and 2872 cm^{-1}

CH_2, two bands at 2926 and 2853 cm^{-1}

CH, one band at 2890 cm^{-1}

(*b*) C—H deformations

CH_3, two bands at 1380 and 1450 cm^{-1}

CH_2, one band at 1465 cm^{-1}

(*c*) $(CH_2)_4$ absorbs at 720 cm^{-1}

(*d*) gem-dimethyl group gives a doublet around 1375 cm^{-1}

ALKENES

(*a*) =C—H stretch : 3000–3100 cm^{-1}

(*b*) C=C stretch   : 1600–1680 cm^{-1}

(*c*) C—H out-of-plane deformation (useful for substitution pattern) : 600–1000 cm^{-1}

(*d*) C—H in-plane deformation (not useful diagnostically) : 1400 cm^{-1} approximately

ALKYNES

(*a*) C—H stretch : 3220–3320 cm^{-1} sharp

(*b*) C≡C stretch : 2100–2300 cm^{-1} weak or absent

(*c*) C—H bend   : 600–700 cm^{-1}

BENZENOID COMPOUNDS

(*a*) 3000–3100 cm^{-1} C—H stretch

(*b*) 1650–2000 cm^{-1} overtone and combination bands (weak).

(*c*) 1550–1600 cm^{-1} ring stretching (very variable intensity)

(*d*) 1450–1500 cm^{-1} ring stretching (very variable intensity)

(*e*) 1000–1300 cm^{-1} C—H in plane bending, usually weak.

(*f*)   600–900 cm^{-1} C—H out-of-plane bending, usually strong.

ALCOHOLS, PHENOLS and ETHERS

(*a*) C—O stretch : 100–1300 cm^{-1}

(*b*) C—O—H bend of variable frequency.

(*c*) O—H stretch can give a lot of information (see Part Five).

AMINES

(*a*) N—H stretch : 3180–3490 cm^{-1}

This is a 100 cm^{-1} doublet for primary amines, a singlet for secondary amines and absent for tertiary amines. It is less affected by hydrogen bonding than the O—H stretch for alcohols, which also absorb in this frequency range.

(*b*) N—H bend 1580–1650 cm^{-1}

ALDEHYDES and KETONES

Ketones

C=O stretch, base frequency : 1719 cm^{-1}

Range : 1660–1780 cm^{-1}

Aldehydes

C=O stretch, base frequency : 1730 cm^{-1}

Range : 1670–1740 cm^{-1}

The position within these ranges is dependent on:

— hydrogen bonding to C=O

— conjugation

— ring size in ketones

C—H str in aldehydes : 2700–2900 cm^{-1}

CARBOXYLIC ACIDS and SALTS

C=O stretch : 1680–1715 cm^{-1}

O—H stretch : 2500–3500 cm^{-1}

C—O stretch : around 1200–1300 cm^{-1}

C—O—H in-plane and out-of-plane bending : around 1400 and 900 cm^{-1} respectively, but variable because of coupling with C—C stretching modes.

ESTERS

C—O stretch : variable but intense, 1000–1400 cm^{-1}

C=O stretch : 1710–1750 cm^{-1}

ACID HALIDES

C=O stretch : highest of all carbonyl frequencies, close to 1800 cm^{-1} or above.

ACID ANHYDRIDES

C=O stretch : doublet (symmetrical and anti-symmetrical), 1730–1850 cm^{-1}

C—O stretch : in fingerprint region, but intense.

AMIDES

C=O stretch : low frequency, 1640–1700 cm^{-1}, varies depending on hydrogen bonding.

N—H stretch : 3100–3500 cm^{-1}, multiplicity can be used to assign primary or secondary amide structures in low polarity solvents.

NITRO-COMPOUNDS

N=O stretch : two bands, 1450–1570 and 1300 to 1370 cm^{-1}

NITRILES

C≡N stretch : sharp band of medium intensity around 2250 cm^{-1}

AZO-COMPOUNDS

N=N stretch : variable intensity around 1570 cm^{-1}

MERCAPTANS

S—H stretch : 2520–2600 cm^{-1}, usually weak, sometimes absent.

C—S stretch : 500–700 cm^{-1}

SULPHOXIDES

S=O stretch : 1000–1100 cm^{-1}, intense.

SULPHONES and SULPHONIC ACIDS

S=O stretch : two bands, 1120–1160 and 1300–1350 cm^{-1}

SULPHONAMIDES

S=O stretch : two bands, 1120–1160 and 1300–1350 cm^{-1}

SULPHONYL CHLORIDES

S=O stretch : two bands, 1150–1210 and 1330–1400 cm^{-1}

HALOGEN-CONTAINING COMPOUNDS

C—I and C—Br stretching frequencies are normally outside the range of commonly used instruments.

C—F stretch : 1000–1400 cm^{-1}

C—Cl stretch : 600–850 cm^{-1}

7.4. IDENTIFICATION OF UNKNOWN COMPOUNDS

You have survived! Now let's apply what you have been studying.

We shall therefore end with a few detailed examples of spectrum interpretation. Recent advances in computer retrieval techniques have extended the range of information available from the infrared instrument by allowing comparison of an unknown spectrum with a bank of spectra of known compounds. This has always been possible in the past by searching libraries of spectra. Once the compound has been classified as an aldehyde or amine etc., the atlas of spectra could be searched for an identical or very similar spectrum, bearing in mind the conditions under which the spectra were obtained. The advent of cheap microcomputers has speeded this search process. The programs work by hunting through stored data to match intensities and frequencies of absorption bands to a given accuracy. The computer will then output the best fits and the names of the compounds found. These can then be examined in an infrared library collection or displayed on a monitor for comparison.

In 1980 the Sadtler data collection on thousands of compounds, became available on floppy disc. Many modern Fourier transform instruments now have these collections available.

We do not have this luxury so here we shall work through a few examples of interpretation of infrared spectra of unknown compounds, trying to get as much structural information as possible.

It is not usually possible by examination of the infrared spectrum of a compound to identify it unequivocally. It is much more normal to use infrared in conjunction with other techniques, eg state, mp, bp, colour, ^{13}C, ^{1}H nmr spectra, mass spectrum, etc. These techniques together can usually give a structural formula for even quite complex organic molecules. Here we shall use only the infrared spectrum and sometimes the molecular formula and/or the relative molecular mass, and as you will see this can be quite powerful.

Let me remind you of the suggested strategy for spectrum interpretation introduced in Section 7.1.

1. Look first at the high-frequency end of the spectrum (above 1500 cm^{-1}) and concentrate initially on the major bands.

2. For each band, short-list the possibilities using a correlation chart (you can find these in the Appendix) and then use more detailed tables if necessary.

3. Use the low-frequency end of the spectrum for confirmation or elaboration of possible structural elements.

4. Do not expect to be able to assign every band in the spectrum.

5. Keep cross-checking wherever possible, eg an aldehyde should absorb near 1730 cm^{-1} *and* in the region 2700–2900 cm^{-1}.

6. Place as much reliance on negative evidence as on positive evidence. Eg if there is no band in the 1600–1850 cm^{-1} region, it is most unlikely that a carbonyl group is present.

7. Band intensities should be treated with some caution: under certain circumstances they may vary considerably for the same group.

8. When looking for small frequency changes be wary; this can be influenced by whether the spectrum was run as a solid or liquid or in solution. If in solution, some bands are solvent sensitive.

9. Do not forget to subtract solvent bands or nujol bands. Do not confuse these with bands from the sample.

I shall work through some examples, thinking out loud to give you a feel of how to go about spectrum interpretation, and then leave you with some problems.

Example 1

Examine Fig. 7.4a. This compound is a liquid and has the molecular formula $C_{10}H_{22}$.

What is the first thing that strikes you?

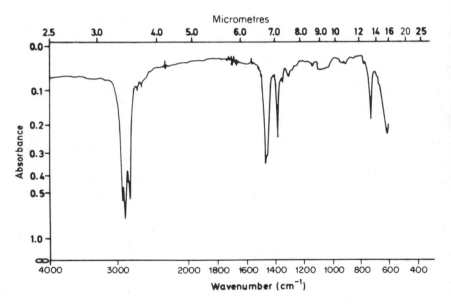

Fig. 7.4a.   *Infrared spectrum of a hydrocarbon, $C_{10}H_{22}$*

Surely, the simplicity of the spectrum, but this is not surprising since this must be a saturated hydrocarbon. We should be able to get a little more information than this however.

The C—H stretching bands confirm that no unsaturation is present, since there are no bands above 3000 cm^{-1}.

The band at 1467 is the scissoring frequency of CH_2 groups and that at 1378 from the symmetrical bending mode of a CH_3 group.

The absence of bands between 1300 and 750 cm^{-1} suggests a straight chain structure, while the band at 782 cm^{-1} tells us that there are four or more CH_2 groups in a chain.

The compound is in fact *n*-decane.

Example 2

Examine Fig. 7.4b. This compound is another liquid with molecular formula C_6H_{14}. What differences are immediately apparent, from the example you have just completed?

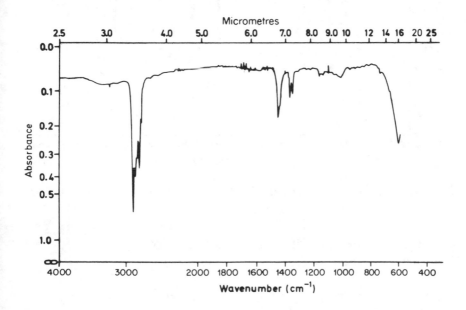

Fig. 7.4b. *Infrared spectrum of an unknown hydrocarbon, C_6H_{14}*

There are two differences. The band in Fig. 7.4a at 1378 cm^{-1} is now split into two bands at 1383 and 1366 cm^{-1}. There must be a gem-dimethyl group present.

Secondly, the weak bands at 1170 and 1145 cm^{-1} confirm this, these are skeletal vibrations from the branched carbon chain.

This is 2-methylpentane.

Example 3

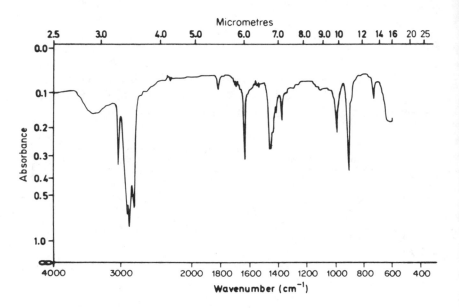

Fig. 7.4c. *Infrared spectrum of an unknown hydrocarbon,* $C_8 H_{16}$

Now examine the spectrum of a compound of molecular formula C_8H_{16}, given in Fig. 7.4c.

Don't ignore formulae, by the way, they can give a lot of information. In this case there must be one C=C double bond in the molecule, since the fully saturated molecule would have a formula C_8H_{18}.

Which peaks in the *spectrum* give us this information?

There are three different sets: the peak above 3000 cm^{-1} tells us that there are hydrogens attached to a double bond, the C=C stretching frequency appears at 1650 cm^{-1}, and finally the C—H deformation bands are at 998 and 915 with a weak peak at 720 cm^{-1}.

What extra structural information do the positions and intensities of these bands give us?

The C=C stretching band is intense and therefore this is not a trans alkene with the double bond far from the end of the chain. The peak at 720 cm^{-1} gives similar information, and tells us we have a chain of at least four CH$_2$s. The C—H deformation bands tell us that the molecule contains the group —CH=CH$_2$.

The only possible structure is therefore oct-1-ene.

Example 4: try this one yourself.

SAQ 7.4a

Oct-1-ene gives the following significant bands in its infrared spectrum. Only the band at 720 cm^{-1} is of weak intensity. (You can check the spectrum in Fig. 7.4c.)

3100 cm^{-1}	C—H str (C=CH$_2$)
2970–2860 cm^{-1}	CH str (alkane)
1650 cm^{-1}	C=C str
998, 915 cm^{-1}	C—H out of plane bending (—CH=CH$_2$)
720 cm^{-1}	—(CH$_2$)$_n$, where $n > 4$

A spectrum of an isomer of oct-1-ene is given in Fig. 7.4d.

Summarise the main differences between the infrared spectra of the two isomers and draw what conclusions you can about the structure of the unknown isomer. $\longrightarrow$

SAQ 7.4a
(cont.)

Before you start you will need a few hints.

The C—H stretching frequency for an alkene varies in the range 3000 to 3100 cm^{-1}.

The band at 840 cm^{-1} has an overtone.

The 'split peaks' just below 1400 cm^{-1} are typical of a tertiary butyl group.

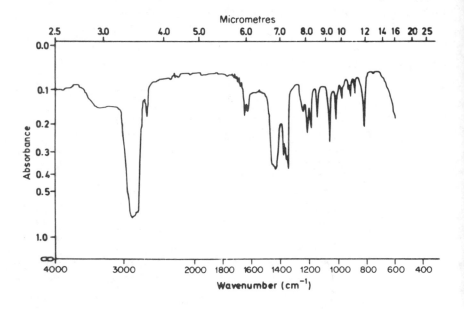

Fig. 7.4d.   *Infrared spectrum of an isomer of Oct-1-ene, C_8H_{16}*

SAQ 7.4a

Example 5

The molecular formula of the compound which gives the infrared spectrum in Fig. 7.4e, is C_8H_8. This tells us that there is the equivalent of five double bonds present. I should point out here that these could be a combination of double bonds, triple bonds or a ring, since a ring also leads to a loss of two hydrogens. Compare hexane, C_6H_{14} and cyclohexane, C_6H_{12}.

One feature which would strike most experienced spectroscopists is the lack of CH_3 and CH_2 deformation bands in the 1300–1500 cm^{-1} range, so there are no alkyl groups in the molecule. There are also no bands just below 3000 cm^{-1}. What does this tell us? The molecule must surely be aromatic.

Typical benzene ring absorptions occur at the following frequencies. Check them off on the spectrum, they are all present.

3000–3100 cm^{-1} (C—H stretch), 1650–2000 cm^{-1} (overtone and combination bands, weak), 1550–1600 cm^{-1} (ring stretching very

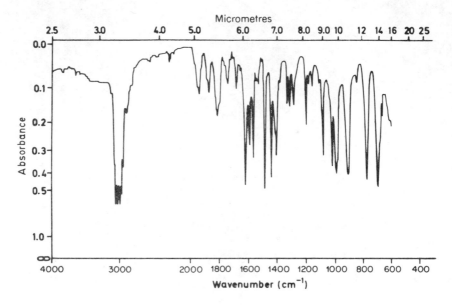

Fig. 7.4e. *Infrared spectrum of an unknown hydrocarbon, C_8H_8*

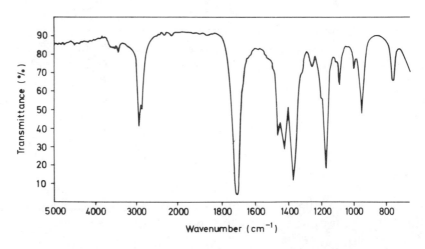

Fig. 7.4f. *Infrared spectrum for an unknown compound, SAQ 7.4b*

variable intensity), 1450–1500 cm^{-1} (ring stretching, very variable intensity), 1000–1300 cm^{-1} (C—H in plane bending, usually weak), 600–900 cm^{-1} (C—H out-of-plane bend, usually strong).

The two peaks at 700 and 780 cm^{-1} indicate a monosubstituted ring. If we subtract C_6H_5 from C_8H_8 we are left with C_2H_3, ie —CH=CH$_2$. The five double bond equivalents can now be seen to come from the double bond (1), the ring (1) and the unsaturation of the benzene ring (3).

Is the presence of a double bond obvious from the spectrum?

The C=C stretch is at 1639 cm^{-1} and the out-of-plane bending frequencies are at 998 and 915 cm^{-1}. All very satisfactory.

Example 6

SAQ 7.4b　　Look at the spectrum in Fig. 7.4f.

(*i*) Write down the frequency of the strongest band and the group you consider to be responsible for it.

(*ii*) There are many types of this group in organic molecules. What types can you rule out from the frequency exhibited here?

SAQ 7.4b

Do you think that this is the spectrum of an aldehyde or carboxylic acid?

Acids have a 'messy' absorption between 2500 and 3500 cm^{-1}, which is absent here. Aldehydes have two C—H stretching bands below 3000 cm^{-1}, again absent here. So we have either a ketone or ester. How can we differentiate between them?

Esters should have a strong band in the region 1000–1300 cm^{-1}. There is no such band present. The band at 1170 cm^{-1} is not sufficiently intense. The C—O stretching frequency is usually as intense as the carbonyl group and certainly always more intense than the CH$_3$ and CH$_2$ deformation bands.

We must therefore have a ketone.

Can we get any more information? Try the following.

SAQ 7.4c

Refer to Fig. 7.4f again and remember we have a ketone.

(*i*)   Is this a saturated or unsaturated molecule?
$\longrightarrow$

SAQ 7.4c
(cont.)

(*ii*) The C—H stretching absorption in this molecule is a doublet. Is this because,

(*a*) one peak is from the C—H stretching mode of the CH_3 group and the other from the CH_2 group?

(*b*) the C—H stretching frequency is split by Fermi resonance with a band at 1400 cm^{-1}?

(*c*) on a low resolution instrument the symmetrical and anti-symmetrical frequencies of the CH_2 and CH_3 groups are superimposed?

(*iii*) Is this a cyclic or open-chain molecule?

Before going further, I would like to expand on the idea introduced above of double bond equivalents in example five. This can be extremely useful in eliminating possible structures if the molecular formula is known.

Double Bond Equivalent (DBE) means any molecular feature that has the same effect on molecular formula as the presence of a double bond. The following are therefore included,

(1) Double bonds, $C=C$, $C=O$, $N=N$, $N=O$ etc.

(2) Rings, each ring is one DBE

(3) Triple bonds $C\equiv C$, $C\equiv N$, each is 2 DBE

The number of DBE's can be calculated using the formula;

$$DBE = 1 + \frac{(2N_c - N_h - N_x + N_n)}{2}$$

where N_c = the number of carbon atoms in the molecule,

N_h = the number of hydrogen atoms in the molecule

N_x = the number of halogen atoms in the molecule

N_n = the number of nitrogen atoms in the molecule

SAQ 7.4d

Calculate the number of DBEs in the following molecules,

(*i*)   Benzene

(*ii*) C_4H_6

(*iii*) $C_7H_{13}NO$

(*iv*) C_4H_3BrO

SAQ 7.4d

The DBEs can be incorporated into structures in various ways. For example, in (*iii*) above we could have two C=C or one C=C and one C=O, or one C=C and one N=O etc. We could have two rings, or one ring and one C=C or C=N and so on.

Let's look at another example where we can apply these ideas.

Example 7

The compound has a molecular formula C_7H_9N

Is it possible for this compound to possess a benzenoid ring?

You can do this using the DBE formula above.

Do this calculation.

I got four DBEs so it is possible and the compound probably contains a benzene ring.

SAQ 7.4e   Interpret the major bands in the spectrum of this compound (Fig. 7.4g), run as a liquid film. If possible suggest a structure or a list of possible structures consistent with this spectrum.

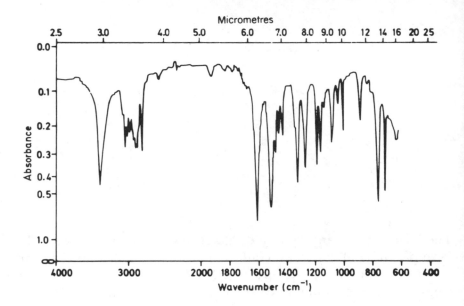

Fig. 7.4g.   *Infrared spectrum of an unknown compound, C_7H_9N*

SAQ 7.4e

Example 8

The spectrum below, Fig 7.4h, was obtained from a 0.5% solution of a liquid sample in a 0.5 mm cell. CCl_4 was employed as the solvent above 1400 cm^{-1} and CS_2 at lower wavenumbers. No solvent peaks are therefore present. The molecular formula of the compound is $C_8H_{10}O$.

How many DBEs does this molecule have?

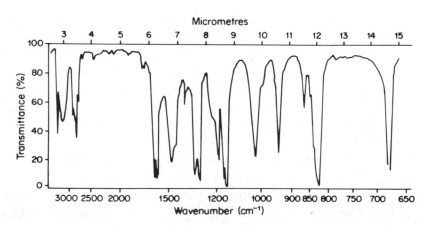

Fig. 7.4h. *Infrared spectrum of an unknown compound, $C_8H_{10}O$ (0.5%) solution*

Using the formula:

$$DBE = 1 + \frac{(2N_c - N_h - N_x + N_n)}{2},$$

$N_c = 8$, $N_h = 10$, hence DBE = 4.

This suggests the presence of an aromatic ring.

∏   List the bands that confirm that this is a substituted benzene.

There are bands both below and above 3000 cm^{-1}, and there are overtone bands in the 1650–2000 cm^{-1} region. There is a doublet at 1600 cm^{-1}, a band at 1470 cm^{-1} and several bands in the C—H out-of-plane and in-plane regions.

There can be little doubt that this is a substituted benzene.

∏   What can we say about the substitution pattern in the ring?

There are three peaks that could be C—H out-of-plane bending vibrations, at 680, 825 and 940 cm^{-1}, the last being outside the normal range. The weak band at 865 is unlikely to be an out-of-plane bend. The possibilities are 1,3,5 or 1,2,4 trisubstitution or 1,3 disubstitution.

∏   List any other functional groups present with the associated bands, that confirm your deduction.

There is an O—H group present from an alcohol or phenol. There are bands from non-hydrogen bonded species at 3500 cm^{-1} and a broad band from polymeric species at 3400 cm^{-1}. The C—O stretching frequency appears at 1140 cm^{-1}.

∏   Suggest possible structures for the compound.

There is difficulty in distinguishing alcohols from phenols. If we assume a tri-substituted benzene and subtract C_6H_3 from the formula we have C_2H_6O left. This could be two methyl groups and an O—H group. In the disubstituted case we could have a methyl group and a CH_2OH group, or an ethyl group and an $O-H$.

I would plump for 3,5-dimethylphenol from the out-of-plane vibrations and a simple chemical test could confirm the presence of a phenolic $O-H$ group.

Example 9

Fig. 7.4i and 7.4j are liquid film spectra of two isomers of molecular formula $C_6H_{10}O_2$. Ignore all bands above 3100 cm^{-1}, these are combination, overtone or impurity bands.

How many DBEs does this molecule possess?

Well,

$$DBE = 1 + \frac{(2N_c - N_h - N_x + N_n)}{2},$$

$N_c = 6$, $N_h = 10$, hence DBE = 2.

This could be a triple bond, two double bonds or two rings or a double bond and a ring.

Is there any evidence of a triple bond?

There are no peaks in the triple bond stretching region, nor a C—H stretching band. This therefore is unlikely.

Is there any evidence of double bonds?

There are two peaks in the double bond region in each spectrum.

In Fig. 7.4i these are at 1724 and 1665 cm^{-1} and in Fig. 7.4j at 1724 and 1642 cm^{-1}.

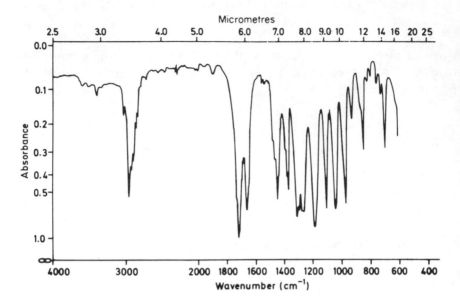

Fig. 7.4i. *Infrared spectrum of an unknown compound,* $C_6H_{10}O_2$

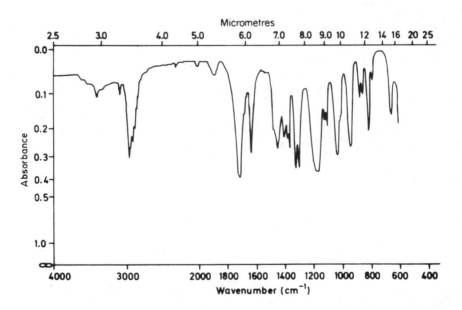

Fig. 7.4j. *Infrared spectrum of an unknown compound,* $C_6H_{10}O_2$

These can easily be assigned as C=O and C=C.

What sort of carbonyl compound is present?

There is no evidence of O—H stretch from a carboxylic acid, yet there are two oxygen atoms in these molecules. Is there a C—O band? I would say the strong bands at approximately 1200 cm^{-1} could be assigned to this absorption.

These must therefore be esters. Furthermore they must be conjugated esters, the ester carbonyl group absorbs at about 1750 cm^{-1} in saturated compounds.

This means we have the molecular fragment,

$$\diagup C = C \diagdown_{CO_2}$$ present

Lots of possibilities still remain, we could have a propyl, ethyl or methyl ester. The following are all possibilities,

H   H   CH$_3$ H   H   CH$_3$
 \   /   \   /   \   /
  C=C   C=C   C=C
 /   \   /   \   /   \
H   CO$_2$C$_3$H$_7$ H   CO$_2$C$_2$H$_5$ H   CO$_2$C$_3$H$_7$

        (a)   (b)   (c)

H   H   CH$_3$ CH$_3$ CH$_3$ H
 \   /   \   /   \   /
  C=C   C=C   C=C
 /   \   /   \   /   \
CH$_3$ CO$_2$C$_2$H$_5$ H   CO$_2$CH$_3$ CH$_3$ CO$_2$CH$_3$

        (d)   (e)   (f)

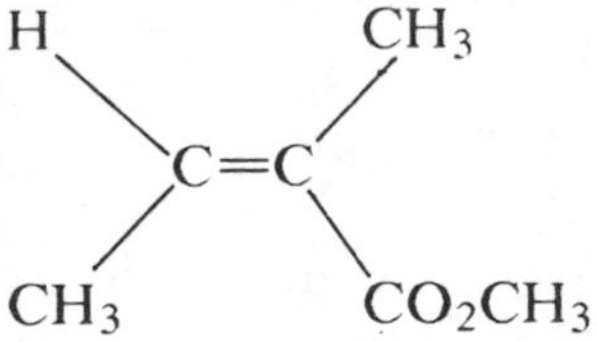

(g) 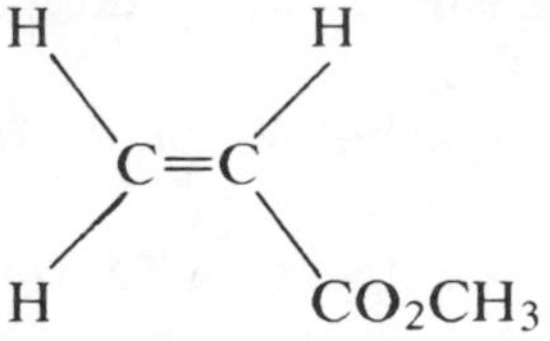   (h)

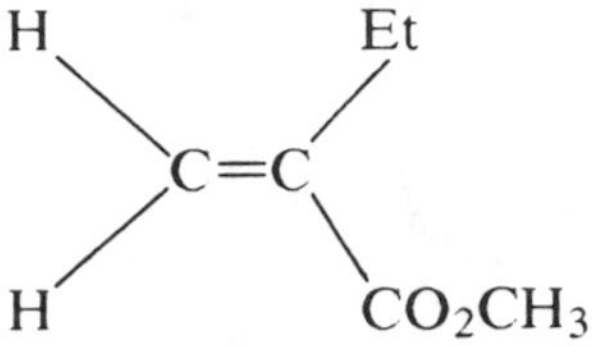

(i) 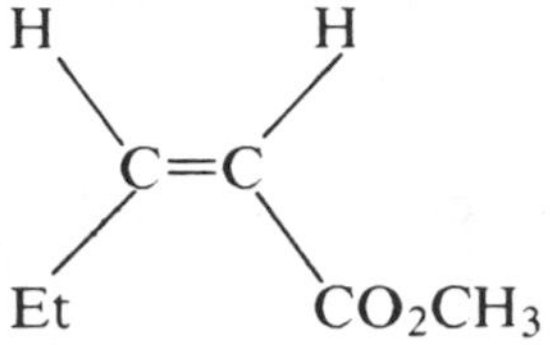   (j)

With the information at your disposal, it is not easy to differentiate these. A study of the C—H deformation region should give information to distinguish cis from trans from vinyl etc. This turns out to be difficult here because of the mesomeric effect of the ester groupings. I can tell you that Fig. 7.4i is the spectrum of cis and trans ethyl crotonate structures (*b*) and (*d*) above, while spectrum Fig. 7.4j is that of ethyl methacrylate, structure (*c*). Try to interpret the C—H deformation region.

Example 10

SAQ 7.4f   Fig. 7.4k is the infrared spectrum of a liquid, of formula C_6H_6NCl.

(*i*)   How many DBEs are present, what does this suggest about the structure? $\longrightarrow$

SAQ 7.4f
(cont.)

(*ii*) Is there evidence for a benzene ring in the
     molecule?

(*iii*) What functional group(s) are present?

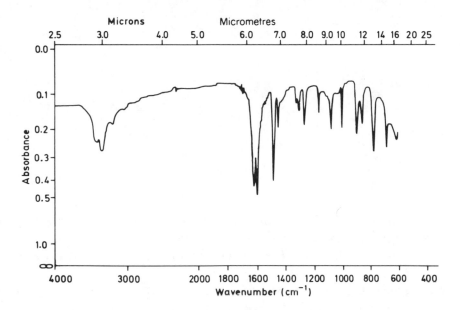

Fig. 7.4k. *Infrared spectrum of an unknown liquid, C_6H_6NCl*

SAQ 7.4f

The compound is 2-chloroaniline, it is difficult to find the C—Cl stretching frequency, a scan to lower wavenumbers is necessary. The two peaks at 680 and 770 cm^{-1} confirm the ortho disubstitution.

In the next section you will find more spectra, I would like you to use the correlation charts, which are reprinted in an Appendix at the end of the Unit, to get as much information as possible from them. The structures of the compounds, with a short commentary on each, are given after the SAQs and responses. This concludes the infrared unit. A lot of material has been presented and I hope you have enjoyed most of it and have found it helpful.

7.5. EXAMPLES FOR FURTHER PRACTICE

This section contains ten spectra. You are asked to glean as much structural information as possible from them.

It is usually not possible to get a complete structure from infrared spectroscopy alone and the technique is usually used in conjunction with simple physical data such as melting point, boiling point, colour and more modern techniques such as mass spectrometry and nuclear magnetic resonance spectroscopy.

You should however be able to assign functional groups in the molecules and classify them as aliphatic, unsaturated, aromatic etc. Use the correlation charts in the Appendix as much as you need.

My interpretation of these spectra are given with the responses to the SAQs at the end of this Unit.

Practice Example, a

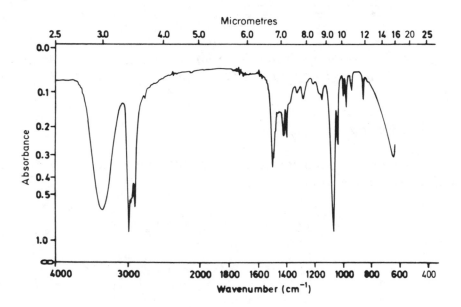

Fig. 7.5a.   *Infrared spectrum of an unknown liquid. [Bp 108 °C, molecular formula $C_4H_{10}O$]*

Practice Example, b

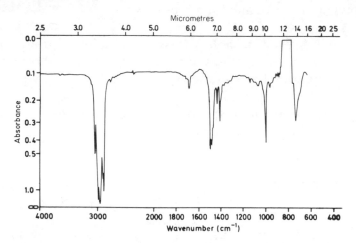

Fig. 7.5b. *Infrared spectrum of an unknown liquid, recorded as a 5% solution in CCl_4 [Bp 122 °C, molecular formula C_8H_{16}]*

Practice Example, c

Two spectra are given for samples at different concentrations in CCl_4. Explain the dilution effects.

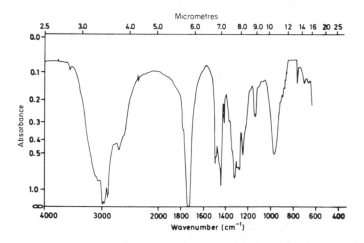

Fig. 7.5c. (i) *Infrared spectrum of an unknown liquid, recorded as a 5% solution in CCl_4. [Bp 202 C, molecular Formula $C_6H_{12}O_2$]*

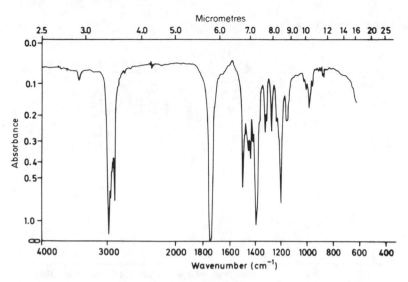

Fig. 7.5c. (ii) *Infrared spectrum of an unknown liquid, recorded as a 1% solution in CCl₄. [Bp 202 C, molecular formula $C_6H_{12}O_2$]*

Practice Example, d

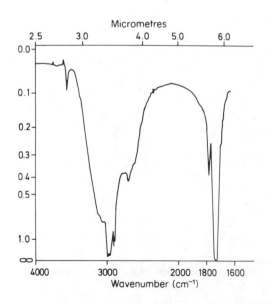

Fig. 7.5d. *Infrared spectrum of an unknown liquid, recorded as a liquid film. [Molecular Formula $C_6H_{12}O$]*

Practice Example, e

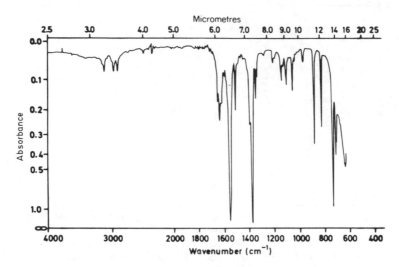

Fig. 7.5e.   *Infrared spectrum of an unknown pale yellow liquid, [Bp 210 C, relative molecular mass 123]*

Practice Example, f.

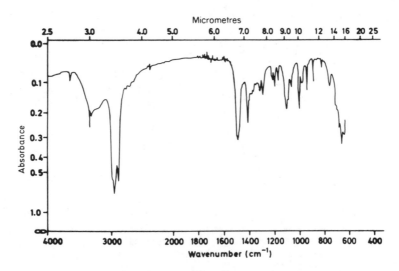

Fig. 7.5f.   *Infrared spectrum of an unknown low melting solid, recorded as a nujol mull. [Molecular formula $C_8H_{12}O$]*

Practice Example, g

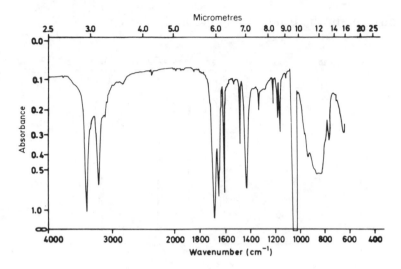

Fig. 7.5g.   *Infrared spectrum of an unknown compound, recorded as a solution in $CHCl_3$. [Molecular formula C_7H_7NO]*

Practice Example, h

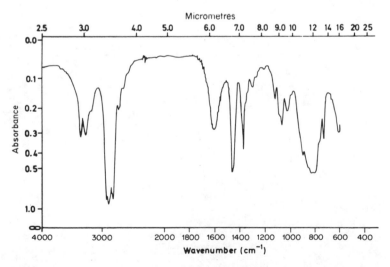

Fig. 7.5h.   *Infrared spectrum of an unknown compound, recorded as a liquid film*

Practice Example, i

Can you explain the doublet at 1740 and 1785 cm^{-1}

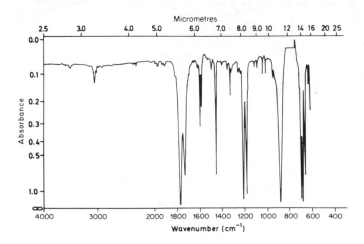

Fig. 7.5i.   *Infrared Spectrum of a compound containing chlorine*

Practice Example, j

Ignore the peak at 2340 cm^{-1}

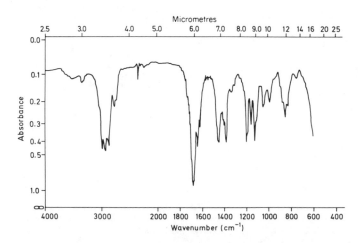

Fig. 7.5j.   *Infrared spectrum of an unknown liquid, [Bp 229 °C, molecular formula $C_{10}H_{16}O$]*

Summary

This Part of the Unit began by dividing the normally studied infrared range into four parts, highlighting and assigning to particular vibrations the bands encountered.

The Part continued with a study of the carbonyl stretching absorption in detail, explaining the small shifts seen in this band in terms of electronic and mechanical effects in molecules.

The ir spectra of the common classes of organic compound were then discussed and a summary of their common absorptions presented.

The Part ended with worked examples and a set of revision spectra for future study.

Student Objectives

On completion of Part Seven you should be able to:

- state the most reliable absorption frequencies for a particular functional group;

- hence ignore many of the bands in infrared spectra as coupled vibrations of variable frequency;

- predict the effect on a group frequency of electronic and mechanical effects and distinguish between them;

- calculate double bond equivalents for a given molecular formula;

- use this information and a correlation chart to distinguish possible structures of unknown compounds;

- have an awareness of modern peak matching search programs for structure determination.

Self Assessment
Questions and Responses

SAQ 1.1a	Give two examples of sources of continuous or polychromatic radiation.

Response

— the sun

— a tungsten bulb filament

What we are looking for here is a source of radiation which produces a continuous spectrum rather than a series of lines. This should have been easy since you were given the answers in the text. But is it the whole truth? Fraunhoffer observed a series of sharp dark lines in the spectrum of the sun. Anybody can see these with quite simple equipment and Newton deserves a reprimand for missing them. Bunsen and Kirchoff working in Heidelberg during the middle of the last century showed that these lines were absorption lines corresponding to emission lines of terrestrial elements. They therefore discovered spectrochemical analysis.

A more accurate answer would be that the inner core of the sun emits a continuous spectrum of polychromatic radiation whereas the outer mantle of the Sun contains elements which absorb specific wavelengths. In this respect the core of the sun behaves like an incandescent solid such as a tungsten filament.

You may have noticed I have begged the question of how wide is the spectral range for which emission is continuous. In this respect radiation from continuous sources is like an old soldier it never dies it only fades away!

Do not worry if these wider aspects of the answer did not occur to you, they serve to introduce an element of lateral thinking which is important for a distance learner as for any other.

****

SAQ 1.1b

Assuming n = 1.68 and 1.64 for violet and red rays respectively sketch the path of sunlight through

(i) a glass prism

(ii) a raindrop.

Response

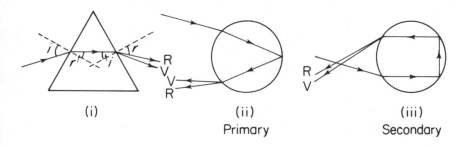

(i) (ii) Primary (iii) Secondary

If you got (i) correct, well done for remembering Fig 1.1a. If you got (ii) or (iii) correct, your optical knowledge of these matters is excellent. This is a revision of an important optical subject which is peripheral to this Section, involving refraction at an interface.

The relationship linking the angle between the incident ray and the normal to a surface, i, and the angle between the refracted ray and the normal, r, is given by Snell's Law.

$$n = \frac{\sin i}{\sin r}$$

If we assume the incident, 'white' radiation is 30° to the normal the angle of the refracted ray to the normal is

Violet $n = 1.68$, $\sin r = \dfrac{\sin i}{n} = \dfrac{0.5}{1.68} = 0.2976$; $r = 17.3°$

Red $n = 1.64$, $\sin r = \dfrac{\sin i}{n} = \dfrac{0.5}{1.64} = 0.3049$; $r = 17.8°$

Hence, in diagrams (i), (ii) and (iii), on entering a medium of higher refractive index the violet light makes a smaller angle with the normal than the red light. A similar argument at the exit surface of the prism shows the violet light is dispersed towards the base of the prism more than the red when projected onto the screen.

****

SAQ 1.1c

> Comment on the sequence of colours in
>
> (i)   the visible spectrum
>
> (ii)   the primary rainbow
>
> (iii) the secondary rainbow.

Response

(i)   The sequence is V.I.B.G.Y.O.R. in decreasing order of angle of refraction.

(*ii*) The observer sees the reflection from those raindrops at the lower edge of the bow as violet and those raindrops at the upper edge of the bow as red.

(*iii*) Because of the second internal reflection the observed order is reversed from (*ii*). Hence violet is at the top edge and red is at the bottom edge.

$$*\!*$$

SAQ 1.1d

The following diagram shows regions of the e.m. spectrum with arbitrary intervals measured in metres.

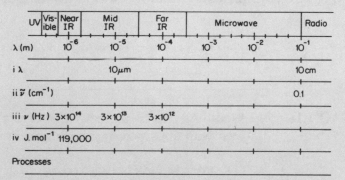

Fig. 1.1d. *The electromagnetic spectrum*

Complete the diagram as follows:

(*i*) enter values in suitable sub-multiples of metres in line i

(*ii*) enter values in wavenumber units in line ii

(*iii*) enter values in frequency units in line iii.

Response

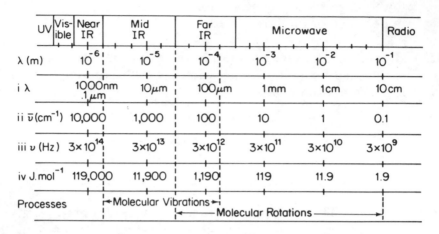

Fig 1.1d (completed).   *Line (iv) will be completed in SAQ 1.1g*

****

SAQ 1.1e   What changes in molecular energies are associated with absorption of radiation in the regions

(*i*)   microwave/far infrared

(*ii*)   mid infrared/far infrared?

Response

(*i*)   molecular rotations

(*ii*)   molecular vibrations

You will find that there is some overlap. Low vibrational frequencies depend on low force constants and high masses. The lowest is at about 50 cm^{-1}. High rotational frequencies depend on low moments of inertia. The highest is about 200 cm^{-1}.

****

SAQ 1.1f	Give two examples of sources of monochromatic radiation.

Response

— radiation from a hydrogen lamp

— radiation from a mercury lamp

The first of these was mentioned in the text and is an example of the electrical excitation of a gas. Any vapour emits line spectra on excitation, and mercury, helium, sodium or neon are common lamp elements. Strictly these emit a series of lines at characteristic wavelength and become a source of monochromatic radiation when one line of very narrow band-width is isolated from the others.

Can you think of any consequences of using one of the above sources for general illumination compared with a normal tungsten filament electric light bulb? Choosing coloured materials in a shop illuminated by one form may cause surprises when the material is inspected at home using another form or in day-light. This raises the topic of colour and its relationship with illumination which is another topic.

****

| SAQ 1.1g | Complete line iv of Fig. 1.1d by calculating the energy in joules per mole for each of the wavelengths listed in the first line of that figure. |

Response

Example for $\lambda = 10^{-6}$

$$\nu = 3 \times 10^{14} \text{ s}^{-1}$$

Using Eq. 1.4, $E = h\nu$

where $h = 66.6 \times 10^{-34}$ Js

$$E = 1.98 \times 10^{-19} \text{ J}$$

So that using $N = 6 \times 10^{23}$ mol^{-1},

the energy per mole will be $E \times N$

$$= 119\,000 \text{ J mol}^{-1}$$

Line iv of Fig. 1.1d has been completed in the response to SAQ 1.1d.

****

SAQ 1.1h

In the experimental arrangement for obtaining an emission spectrum (Fig. 1.1e) the sample itself emits radiation in 6 spectral regions. This is therefore, a lime source since a series of frequencies (lines) are emitted with many regions of the spectrum devoid of radiation.

In the case of the arrangement for obtaining an absorption spectrum we observed the following.

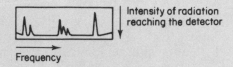

If we remove the sample no absorption of radiation takes place, and we are left with the emission spectrum of the source (a continuous source).

Draw the emission spectrum of this source; it is called the background.

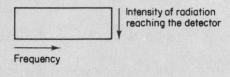

Response

Response

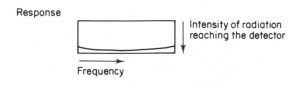

****

SAQ 1.2a   From Fig. 1.1d complete the following

(*i*)   The infrared/microwave interface is
        m

(*ii*) The infrared/microwave interface is
        cm^{-1}

(*iii*) The region indicated for molecular vibra-
        tions is from   to   m, or
        from   to cm^{-1}

(*iv*) The region indicated for molecular rota-
        tions is from   to   m, or
        from   to   Hz.

Response

(*i*) 5×10^{-4} m

(*ii*) 20 cm^{-1}

(*iii*) 2.5×10^{-6} to 2.5×10^{-4} m, or from 4000, 40 cm^{-1}

(*iv*) 5×10^{-5} to 10^{-1} m, or from 1.2×10^{13} to 3×10^{9} Hz

****

SAQ 1.2b   Assuming the equilibrium frequency of the
                $H^{35}Cl$ molecule corresponds to 2886 cm^{-1} cal-
                culate the force constant of the $H^{35}Cl$ molecule
                in units N m^{-1} using the integer isotopic relative
                atomic mass value A_r (^1H) = 1, A_r (^{35}Cl) = 35.

Response

$$H^{35}Cl \ \mu = \frac{H \times {}^{35}Cl}{(H + {}^{35}Cl)N} = \frac{35}{36 \times 6.022 \times 10^{23}}$$

$$= 1.614 \times 10^{-24} g$$

$$= 1.614 \times 10^{-27} \ kg$$

$$f = 4\pi^2\nu^2\mu = 4\pi^2 (2886 \times 2.998 \times 10^{10})^2 \times 1.614 \times 10^{-27}$$

$$= 477 \ N \ m^{-1}$$

****

SAQ 1.2c   Assuming the equilibrium frequency of ${}^{12}C^{16}O$ corresponds to 2143 cm^{-1} calculate the equilibrium frequency of ${}^{13}C^{16}O$ using integer mass numbers.

Response

$$\frac{\nu_{13-16}}{\nu_{12-16}} = \frac{\sqrt{\frac{29}{13 \times 16}}}{\sqrt{\frac{28}{12 \times 16}}} = \sqrt{\frac{29 \times 12 \times 16}{28 \times 13 \times 16}} = 0.97778$$

$$\nu_{13-16} = 0.97778 \times 2143 = 2095 \ cm^{-1}$$

****

SAQ 1.2d   Which of the molecules listed below absorb infrared radiation?

H_2, HD, D_2, HF, F_2.

Response

Only HF has an appreciable dipole moment and hence a strong infrared absorption spectrum. If you included HD you were strictly correct because there is a very small dipole moment arising from the slightly different properties of the two isotopes but this would only lead to a very weak absorption.

****

SAQ 1.2e

List the following molecules in decreasing order of equilibrium vibration frequency

IBr, HI, IF, ICl.

Response

The respective wavenumbers are HI 2230 cm^{-1}, IF 604 cm^{-1}, ICl 381 cm^{-1}, IBr 267 cm^{-1}. Here the force constants are fairly similar and the equilibrium frequency is largely determined by the reduced mass. In simple terms think of I as a fixed large mass and H, F, Cl and Br as increasing masses vibrating at decreasing frequencies.

****

SAQ 1.2f

List the following in increasing order of number of probable fundamental absorption bands in the infrared

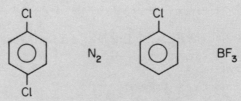

N_2 BF_3

Response

Response

		Cl	Cl
N_2	BF_3	(benzene ring with Cl)	(benzene ring with Cl top and Cl bottom)
(1)	(2)	(3)	(4)

The larger the number of atoms the more vibrations and more absorption bands may be expected. However there is the activity of vibrations to be considered and a molecule with particular symmetry properties may have a number of inactive vibrations. In this list p-dichlorobenzene is in this category since it has a centre of symmetry as well as several planes of symmetry and axes of symmetry. The relationship between symmetry and vibrational activity is a large subject and is outside the scope of this Unit.

****

SAQ 1.2g   Tick the viable combinations for spectral activity in the following table.

	Permanent dipole		No permanent dipole	
	Vibrat.	Rotat.	Vibrat.	Rotat.
Solid				
Liquid				
Gas				

Response

	Permanent dipole		No permanent dipole	
	Vibrat.	Rotat.	Vibrat.	Rotat.
Solid	✓		✓	
Liquid	✓		✓	
Gas	✓	✓	✓	

****

SAQ 1.2h   Sketch the potential function representing harmonic oscillation for a molecule with a low force
constant and one with a high force constant. Suggest a typical pair of molecules in this category.

Response

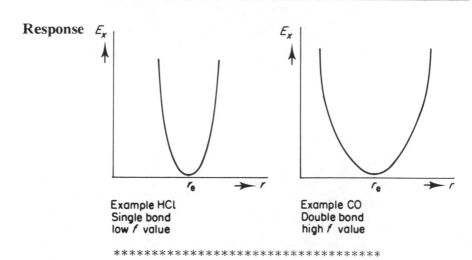

Example HCl
Single bond
low f value

Example CO
Double bond
high f value

****

SAQ 1.3a | Compare the infrared spectrum of CO as a gas (Fig. 1.3b) and as a solution in CCl4 (Fig. 1.3c). What does this suggest about the motion of the CO molecules in the two states?

Response

CO rotates freely in the gas phase but not in solution in CCl_4. Intermolecular interactions cause line broadening.

****

SAQ 1.3b | Fig. 1.3a and Fig. 1.3b shows the comparison of a portion of the spectrum of CO recorded on two different infrared spectrometers. What do you conclude about the difference in the operating condition of the two instruments?

Response

The two spectrometers are operating at different resolutions.

There is a common trap here of comparing like with like. Always be careful to compare pairs of observations in which there are controlled common elements and different elements. For example, when comparing the same sample using different spectrometers as in this case, or different samples in the same spectrometer. The first pair of results tells you something comparative about the spectrometers; the second pair of results tells you something comparative about the samples.

****

SAQ 1.3c   A typical energy change associated with a rotation transition is 15×10^{-23} J.

For a vibration transition it will be in the region of 100 times greater.

Calculate the energy in joules, of the vibration transition 2143 cm^{-1} for a CO molecule.

Response

Use Eq. 1.12, $\nu = \dfrac{E_1 - E_0}{h}$

$$\nu = \frac{c}{\lambda}$$

$$= \frac{3 \times 10^{10} \text{ cm s}^{-1}}{1/2143 \text{ cm}}$$

$$= 6.429 \times 10^{13} \text{ s}^{-1}$$

$$E_1 - E_0 = 6.429 \times 10^{13} \times 6.6262 \times 10^{-34}$$

$$= 42.6 \times 10^{-21} \text{ J}$$

Note that the use of cm^{-1} is often used as a measure of energy. It provides a convenient link between wavenumber values of spectral bands and differences in molecular energy levels.

Hence a band at 4286 cm^{-1} would be associated with an energy change of 85.2×10^{-21} J.

A typical energy change of 15×10^{-23} J could be observed at 7.5 cm^{-1}.

****

SAQ 1.4a

The table below lists data for the relative population of two energy states E_0 and E_1 for various temperatures and for energy separations represented by spectral transitions in wavenumbers, cm^{-1}.

T/K $\bar{\nu}/cm^{-1}$	10	30	100	300	1000	3000
10	0.24		6×10^{-7}	0	0	0
100	0.87	0.65	0.24		6×10^{-7}	0
300	0.96	0.87	0.62	0.24	8×10^{-3}	6×10^{-7}
1000	0.99	0.96	0.87	0.65	0.24	

(*i*) Three gaps have been left for you to complete; calculate the relative population of the higher to the lower level assuming the degeneracy of each level is the same. You might like to use the same equation that I used in calculating the listed values:

$$\frac{N_1}{N_0} = \exp\left(\frac{-(E_1 - E_0)}{kT}\right) =$$

$$\exp\left(\frac{-hc\bar{\nu}_{1\leftarrow0}}{kT}\right) = \exp\left[-1.438\left(\frac{\nu_{1\leftarrow0}}{T}\right)\right]$$

(*ii*) Does the population of the excited state E_1, increase or decrease as the temperature is raised?

(*iii*) Is the population of the excited state greater or smaller for the 1000 cm^{-1} transitions compared with those at 30 cm^{-1}, at a given temperature? $\longrightarrow$

SAQ 1.4a
(cont.)

(*iv*) How do you think the intensity of a transition of the type $E_2 \leftarrow E_1$ will differ for the two cases:

$$T = 100 \text{ K}, E_1 \leftarrow E_0 = 10 \text{ cm}^{-1}$$

$$T = 100 \text{ K}, E_1 \leftarrow E_0 = 300 \text{ cm}^{-1}?$$

Response

(*i*)   0.01 in all three cases.

(*ii*) increase: The ratio increases as you go down any column

(*iii*) smaller: The ratio decreases as you go along any row, left to right.

(*iv*) To a first approximation the intensity of the transition depends upon the population of the first excited state.

When $T = 100$ K and $E_1 \leftarrow E_0 = 10 \text{ cm}^{-1}$ the population of E_1 is fairly high, $\dfrac{N_1}{N_0} = 0.87$.

When $T = 100$ K and $E_1 \leftarrow E_0 = 300 \text{ cm}^{-1}$ the population of E_1 is fairly low, $\dfrac{N_1}{N_0} = 0.01$.

It is obvious that the initial population of the E_1 state in the second case is much less than in the first case, and this will reflect significant differences in intensity of $E_2 \leftarrow E_1$ transitions, all other factors being equal.

****

SAQ 1.5a How many vibrational degrees of freedom do the following molecules possess?

(i) Ammonia (NH_3) Answer ...

(ii) Ethene ($CH_2=CH_2$) Answer ...

(iii) Propyne ($CH_3C\equiv CH$) Answer ...

Response

The numbers are entered on the basis of the $3N - 6$ rule for $N = 4, 6$ and 7 respectively.

(i) 6

(ii) 12

(iii) 15

For some purposes the CH_3 group is taken as one unit and the internal C—H vibrations treated separately. In this approximation the molecule becomes linear ($XC=CH$ where $X = CH_3$) and we apply the $3N - 5$ rule for $N = 4$ leading to 7 modes of vibration. This leaves 6 internal CH_3 vibrations including the torsion which occurs a low wavenumber (approximately 100 cm^{-1}).

****

SAQ 1.5b Can you think of examples of 'resonance' in the areas of civil engineering, electrical engineering or acoustics?

Response

There are examples, some apocryphal, of bridges collapsing because natural frequencies of vibrations have been induced by columns of marching soldiers. For this reason marchers may be required to break step. There are also spectacular records of large suspension bridges vibrating in strong winds with amplitudes increasing to destruction, I included an ambitious SAQ on this topic to encourage some wider vision!

Radio is of course a form of spectroscopy. The e.m. signal from a transmitter is detected by an aerial and when the circuit in the receiver is tuned to the correct frequency the signal is amplified and converted into an audio signal at the loud speaker.

Acoustically wine glasses can be shattered by notes of a suitable frequency. A more controlled experiment is possible with a set of tuning forks. A vibration from one can be transmitted through air or some other suitable medium to another providing the receiver has the same natural frequency as the transmitter.

****

SAQ 1.5c

List the characteristic group frequencies in the following with reference to Fig. 1.5e.

HCN ...

ClCN ...

To what do you attribute any characteristic features of these values?

Response

| HCN | C≡N | = 2089 cm^{-1} | C—H | = 3312 cm^{-1} |
| ClCN | C≡N | = 2201 cm^{-1} | C—Cl | = 729 cm^{-1} |

Characteristic features are attributed to the large force constant of the C≡N bond and the small and large mass effect of the H and Cl atoms respectively.

Each molecule has two stretching frequencies. In the case of HCN, ν_1 is shown as essentially C≡N stretching and for ClCN, ν_3 is also shown as essentially C≡N. The wavenumber values are in the region 2000 cm^{-1} to 2250 cm^{-1} which is found to be the characteristic region for all nitriles. The remaining stretching vibration is assigned at 3312 cm^{-1} in HCN and 729 cm^{-1} in ClCN. The modes are shown to be largely C—H and C—Cl stretching motions and the spectral regions are again found to be characteristic for these groups. Wavenumber values are determined by force constants of bonds and masses of atoms. From the form of the equation for a diatomic molecule,

$$\nu_e = \frac{1}{2\pi}\sqrt{\frac{f}{\mu}}$$

atoms of light mass such as H vibrate at high wavenumber values. Hence all C—H groups give rise to characteristic wavenumber values in the 2300–3300 cm^{-1} region. By contrast bonds involving atoms of heavy mass such as Cl vibrate at low wavenumber values and all compounds containing C—Cl groups have strong bands in the 700–800 cm^{-1} region.

Multiple bonds can give rise to characteristic wavenumber values because of the larger values of the force constants. Thus double bonds (C=C, C=N, C=O) can absorb in the 1600–1850 cm^{-1} region and triple bonds (C≡C, C≡N) can absorb in the 2000–2300 cm^{-1} region. The intensity considerations which determine whether the associated vibrations are inactive, weak or strong will be considered in Part 4 of this Unit.

$$************************************$$

SAQ 1.5d   Suggest the likely forms of the modes of vibrations of the formate ion. It will help to consider this system as an extension of a non-linear triatomic molecule which is illustrated in Fig. 1.5c.

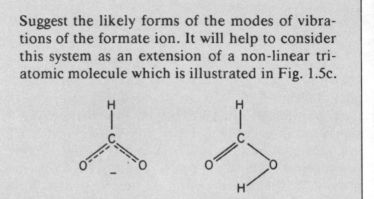

Formate ion   Formic acid

Response

Firstly you should establish that the formate ion has 6 modes, since there are four atoms. $N = 4$, hence $3N - 6 = 6$ modes. Of these, three will correspond to those in non-linear triatomics (Fig. 1.5c and d) namely:

1. antisymmetric stretching of the $C-O$,
2. symmetric stretching of the $C-O$,
3. bending of the $C-O$.

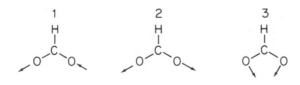

The remaining three are likely to be:

4. C—H stretch,
5. C—H in-plane bend,
6. C—H out-of-plane bend

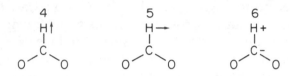

This is an interesting example of structure determination because it assumes that the two C—O bonds are equivalent and that the negative charge is equally shared by the two O atoms. An alternative model on which to base assumptions would be one in which one O was double bonded to the C and the other O was single bonded to the C. The two bonds would no longer interact to give equivalent symmetric and antisymmetric vibrations but would yield a C=O stretching vibration and a C—O stretching vibration. It is found that spectroscopic data is consistent with the former model rather than the latter.

****

SAQ 1.5e

Predict the main features of the infrared spectrum of formic acid.

Response

This question is open ended and is also influenced by prior knowledge. Do not be surprised if your answer is wildly different from mine!

We have noted from a question asked in Section 1.5.1 that an OH group gives rise to an absorption band near 3600 cm^{-1} when the sample is in the vapour state. This value is reduced for samples in the condensed state and is heavily dependent on hydrogen bonding. The observed spectrum for the OH group is therefore strongly dependent on the phase.

We also noted in another question in Section 1.5.2, that C=O groups lead to characteristically strong absorption bands in the 1600–1850 cm^{-1} region. This value may also be reduced by hydrogen bonding (see Part 5 of this Unit).

It is a feature of carboxylic acids, of which formic acid is the simplest, that strong dimeric forms occur via OH—O=C bonds which are only broken in dilute solution or at low vapour pressures.

What about the rest of the spectrum? Since $N = 5$ we expect $3(5) - 6 = 9$ fundamental modes. We have accounted for two characteristic group frequencies. The three associated with the C—H group will be similar to those of the formate ion which we considered in SAQ 1.5d. The remaining four fundamental modes will be skeletal modes associated with C—O groups and OH bending motions.

****

SAQ 1.5f	List five diatomic molecules which are infrared active and five which are infrared inactive.

Response

You could select from the following common materials:

ir active

HF, HCl, HBr, HI, BrCl, IBr, KCl, LiCl, etc.

ir inactive

H_2, O_2, N_2, Na_2, Br_2, Cl_2, F_2, I_2, etc.

****

SAQ 1.5g	State the rule governing the activity of any fundamental vibrational mode in the infrared region.

Response

You may recall from the text that in qualitative terms a vibrational mode is active in the infrared if there is a change in dipole moment during the vibration.

When it is not feasible to judge this criterion symmetry raises its head.

****

SAQ 1.5h	Indicate the factors influencing the intensities of absorption bands of C=O and C=C stretching modes.

Response

You can either consider the magnitude of the change in dipole moment or the symmetry of the vibration. Both approaches say the $C=O$ and $C=C$ stretching vibrations correspond to strong and weak bands respectively.

The $C=O$ group has its own built in dipole moment and asymmetry. Whatever the substituents other than an opposing $C=O$ group as in carbon dioxide, carbon suboxide, or some inorganic carbonyls, the band will always be strong and has a characteristic group frequency in the 1600–1850 cm^{-1} region.

By contrast you should have noticed that the activity of the $C=C$ group depends on the nature of the four substituents. If they form a symmetrical system the vibration will be inactive in the infrared. If they form an unsymmetrical system the vibration will be active. Accordingly $C=C$ stretching vibrations may be absent, (as in ethene or *trans* dichloroethene) or weak (as in *cis* dichloroethene) or medium (as in 1,1-dichloroethene).

Hence force constants, mass and group frequency considerations determine where an infrared band may occur, but dipole moment and symmetry considerations decide if you see it!

****

SAQ 2.1a

Complete the following table with ticks for compatible combinations and crosses for incompatible combinations.

	Constant Resolution	Constant Energy
Constant Slit		
Programmed Slit		

Response

The following is generally the case for prism instruments but many grating instruments offer fairly constant resolution over a considerable region for a constant slit:

	Constant Resolution	Constant Energy
Constant Slit	×	×
Programmed Slit	×	√

It is common practice to programme the slits to achieve constant energy. At wavenumber values where the energy of the source is high or low the slits are automatically changed to narrow or wide respectively to achieve a constant signal.

The resolution obtained depends upon the slit-width and resolving power of the dispersive element at that particular wavenumber value. Only if the latter were constant would the resolution be constant at constant slit. The energy reaching the detector also influences resolution since the signal must be readily measureable in a reasonable time period.

****

SAQ 2.1b	(i) Sketch a single beam infrared spectrum of any sample for which you can recall the general appearance of the spectrum. What about CO?
	(ii) Sketch a double-beam infrared spectrum of the same sample.

Response

(*i*)

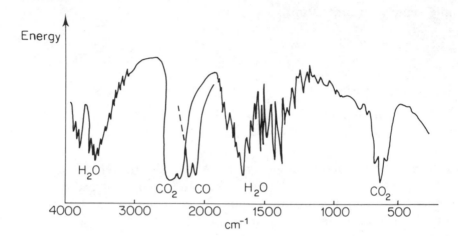

Here the absorption of any particular sample is added to that of atmospheric absorption.

(*ii*)

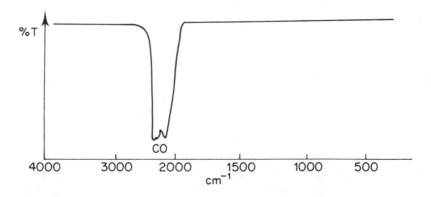

This is the form of the spectrum which is normally displayed using a double beam instrument. There is a trap here which many fall into unless they are aware of the appearance of the single beam result. Firstly there are regions where atmospheric absorption occurs in which any sample measurement is suspect. Secondly there are regions where the spectrometer is energy limited and measurements of absorbance are less accurate than the regions where higher energy are indicated in the single beam spectrum.

Are you convinced that it is worth knowing something about the appearance of a single beam spectrum?

****

SAQ 2.1c	Complete the following table by entering ticks for compatible combinations and crosses otherwise.

	Dynamic Range		
	10–99%T	1–99%T	0.1–99.9%T
Ft-ir	√	√	√
Ratio Recording			
Servo Mechanism			

Response

The following are typical performance indicators:

| | Dynamic Range | | |
	10–99%T	1–99%T	0.1–99.9%T
Ft-ir	√	√	√
Ratio Recording	√	√	×
Servo Mechanism	√	×	×

****

SAQ 2.1d

List common materials for prisms.

Comment on the relationship between relative atomic mass of the elements in the prism and the cut-off point to infrared radiation.

Response

Common materials are NaCl and KBr with cut-offs at about 650 cm^{-1} and 400 cm^{-1} respectively. The relative atomic masses are 23–35 and 39–80. Since vibrational frequency is inversely proportional to the square root of the reduced mass the latter will have the lower wavenumber value.

Less common prism materials are LiF and CsI which have low wavenumber cut-offs at 1000 and 200 cm^{-1} respectively for the reasons outlined above. The dispersion, and therefore the resolution which may be achieved, is greatest near to the cut off. This is because the refractive index changes sharply with wavenumber near the cut-off. The resolution near 3000 cm^{-1} is therefore poor for NaCl, KBr and CsI but good for LiF. Before the advent of routine grating or Ft-ir spectrometers LiF prism instruments were used to study X—H stretching vibrations.

****

SAQ 2.1e

Sketch the simplest arrangement you can for obtaining a dispersed spectrum and scanning this across an exit slit.

Response

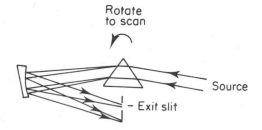

A common variant to this is the monochromator system shown in Fig. 2.1a in which the dispersed radiation is reflected back through the prism by the Littrow mirror which may be rotated to cause the spectrum to scan across the exit slit.

****

> **SAQ 2.1f**   In terms of the trading rules suggest ways in which the following may be achieved:
>
> (*i*)   High speed of recording a spectrum.
>
> (*ii*) High speed of displaying a spectrum.
>
> (*iii*) High resolution with no time restraints.
>
> (*iv*) High resolution with severe time restraints.
>
> (*v*)   Low signal to noise with ability to rapidly record a spectrum.

Response

(*i*)   Wide slit and high gain.

(*ii*) As (*i*) since a spectrum from a dispersive instrument is normally displayed almost instantly.

(*iii*) Narrow slit and moderate gain.

(*iv*) Narrow slit and high gain.

(*v*)   Wide slit and moderate gain.

These illustrate 'the trading rules' which indicate that in spectroscopy you cannot get anything for nothing! The first generation of instruments required much skill of an operator to select the parameters to suit his or her needs. Modern instruments frequently allow the operator to select a particular parameter such as resolution, signal to noise or scan time and then pre-determine limits of the remaining parameters. Nevertheless there will be times when you, the operator will need to understand the principles under which the instrument operates to achieve the best results.

****

SAQ 2.2a

Compare the advantages and disadvantages of Ft and dispersive infrared spectroscopy by completion of the table below. I have started you off by indicating that Ft-ir has the advantage of speed.

	Dispersive ir	Ft-ir
Cost		
Resolution		
Speed		√
Computer access		
Signal/Noise		
Routine samples		
Intractable samples		
Wide cm^{-1} regions		
Narrow cm^{-1} regions		
Precision in cm^{-1}		
Accuracy in cm^{-1}		
Low stray light		
Double beam work		

Response

There are subjective value judgements to be made and the answers change with technical advance but reasonable choices are as follows

	Dispersive ir	Ft-ir
Cost	√	
Resolution		√
Speed		√
Computer access		√
Signal/Noise		√
Routine samples	√	
Intractable samples		√
Wide cm^{-1} regions		√
Narrow cm^{-1} regions	√	
Precision in cm^{-1}		√
Accuracy in cm^{-1}		√
Low stray light		√
Double beam work	√	

****

SAQ 2.2b

Radiant energy is measured as a function of wavenumber, $\bar{\nu}$ by using a dispersive spectrometer.

Radiant energy is measured as a function of difference in path length, δ, using an Ft-ir spectrometer.

(*i*)   What is the name given to the pair of equations linking the two spectra resulting from these measurements?

(*ii*) What limits the wavenumber range of a spectrometer?

(*iii*) What limits the useful maximum δ that is employed in Ft-ir spectrometers?

Response

(*i*) Fourier transform pair

(*ii*) The high and low limits of $\bar{\nu}$ are determined by the ability of the optical materials to transmit radiation and the ability of the detection and measurement system to register it on the read-out device. These factors determine the range of the spectrometer which is typically 400–4000 cm^{-1}. The resolution at any point in this range, which is typically 1 cm^{-1}, is also determined by optical properties of the system and particularly of the dispersive element.

(*iii*) The maximum value is determined by the ability to move the mirror and measure the movement accurately.

Note that the distance of travel of the moving mirror above and below the position of the fixed mirror is $\delta/2$. The limits of $+\delta/2$ and $-\delta/2$ determine the path difference δ and hence the resolution of the Ft-ir instrument.

$$*******************************$$

SAQ 2.2c

(*i*) Sketch the optical arrangement for a Michelson interferometer.

(*ii*) Sketch an interferogram for a monochromatic source.

(*iii*) Sketch an interferogram for a polychromatic source.

(*iv*) In order to take an absorption spectrum of a sample where should the sample be located in the beam?

Response

Typical sketches and location are as follows

(*i*)

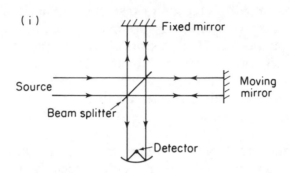

(*ii*) and (*iii*)

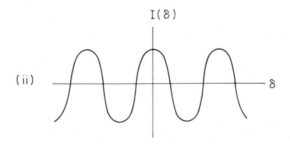

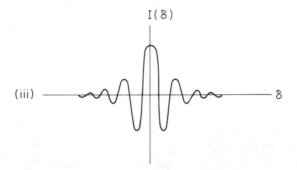

(*iv*) The sample is normally located immediately in front of the detector.

****

SAQ 2.2d

Complete the table below, by inserting two ticks appropriate to matching together the pairs of advantages.

	Jacquinot Advantage	Fellgett Advantage
Multiplex advantage		
Throughput advantage		

Response

	Jacquinot Advantage	Fellgett Advantage
Multiplex advantage		√
Throughput advantage	√	

****

SAQ 2.2e	Briefly define the following terms in relation to Ft-ir:
	Apodisation
	Digitisation
	Phase Correction
	Signal Averaging

Response

Apodisation – The removal of the side lobes which are either instrumental or computational artifacts on a spectral band generated from an interferogram by a Fourier transform.

Digitisation – The conversion of a continuous analogue signal into a discontinous digital signal which is a necessary stage in the Ft-ir instrument.

Phase correction – This is the correction which is made in order to ensure that the first reading is taken at $\delta = 0$ and that subsequent readings are taken at equal intervals on either side of the first.

Signal averaging – Ft-ir spectrometers, like other rapid scanning instruments, can improve signal to noise by repeat scanning so that systematic real signals are accumulated but random fluctuations such as noise average out. It is a general principle that the improvement of signal to noise is proportional to the square root of the number of scans. The limitations of the method are ability to scan very quickly, ability of the instrument to reproduce measurements and amount of instrument time available.

****

SAQ 2.2f | On lifting the lid of a Fourier transform spectrometer two sources of radiation were observed – a ceramic filament and a He/Ne laser. What are their functions?

Response

The ceramic filament is the source of continuous radiation from which an absorption spectrum is obtained. The He–Ne laser generates interference fringes which are used to measure the position of the moving mirror.

****

SAQ 2.2g | An Ft-ir spectrometer is used to record a single beam spectrum from a single scan and a difference in path-length, δ, of 100 mm. A noise level of 1% is recorded, the S/N being 100/1.

(*i*) How many scans are needed to improve the noise to 0.5% and to 0.01%?

(*ii*) What is the limiting resolution, in units of cm^{-1}?

(*iii*) How could a limiting resolution of 0.02 cm^{-1} be achieved?

Response

(*i*) $S/N \propto \sqrt{n}$

For $n = 1$, $S/N = 100/1$

to improve to a $S/N = 100/0.5$, $\sqrt{n} = 2$ and $n = 4$

to improve to a $S/N = 100/0.1$, $\sqrt{n} = 10$ and $n = 100$

(ii) $\delta = 100$ mm $= 10$ cm

Limiting resolution $= 1/10 = 0.1$ cm^{-1}

(iii) To obtain a limiting resolution of 0.02 cm^{-1} g, $\delta = 1/0.02 = 50$ cm $= 500$ mm

****

SAQ 2.3a

> For any computer with which you may be familiar estimate or look up the following
>
> Memory Size Bytes K
>
> Read only ROM
>
> Random Access RAM
>
> Backup

Response

In my own case I use three microcomputers which are interfaced, or capable of interface to a spectrometer:

(i)   The BBC Model B.

(ii) The Perkin–Elmer Model 3600 Data Station.

(iii) The Digilab FTS 50 Data System.

Many other systems exist but I will list the parameters for the three I know in the spirit of the question!

| | Memory Size Bytes K | | |
	(i)	*(ii)*	*(iii)*
Read only ROM	16 K	64 K ⎫	200 K
Random access RAM	32 K	⎬	1000 K
backup	160 K/Disc	⎭ 160 K/Disc	160 K/Disc Winchester 19,000 K

****

SAQ 2.3b   List the four main components of a computer.

Response

1. An input device such as a key-board or a spectrometer on-line.

2. A memory which may be in various forms such as permanent, known also as read-only-memory; volatile memory known also as random access memory; back-up memory such as disc or tape with which information may be exchanged with the computer; screen memory in which final information is held.

3. The central processor unit which accepts instructions from the program which may be loaded into the volatile memory or held in permanent memory and operates those instructions on the data which is entered.

4. The output device which is typically a screen on which output is

assembled and suitably formatted for print-out as 'hard-copy' in the most desired form.

****

SAQ 2.3c

A double sided floppy disc holds 80 K on each side, and a Winchester disc holds 25000 K of memory. Assuming an infrared spectrum requires 7 K at 1 data point per wavenumber between 600 and 3600 cm^{-1},

(*i*) How many spectra can be stored on the Winchester disc?

(*ii*) How many floppy discs would you buy to store 500 spectra?

(*iii*) If spectra were run between 2500 and 3500 cm^{-1} at 0.05 data points per wavenumber how many spectra could now be stored on a floppy disc?

Response

(*i*)   25 000/7 = 3571.4   Answer 3571 spectra

(*ii*) 1 Disc takes 160 K

160/7 = 22.85; 1 Disc takes 22 spectra

500 spectra require 500/22 = 22.72 Discs. Answer 23 Discs

(*iii*) 600–3600 cm^{-1} at 1.0 cm^{-1} data points which is 3000 data points enables 22 spectra to be stored on disc

2500–3500 cm^{-1} at 0.05 cm^{-1} per data point is 1000/0.05 = 20 000 data points.

Reduction in spectra is 22 × 3000/20 000 = 3.3   Answer 3 spectra

****

SAQ 2.3d	List twelve computer operations in decreasing importance (in your judgement and experience) for interfacing purposes with infrared spectrometers.

Response

This is also an open ended question but my own experience would be for the following operations in decreasing priorities:

Save, Load, Enhance weak bands, Subtractions, Peak heights, Accumulate, Control, Format, Enhance overlapped bands, Search.

No doubt your own priorities will differ but it is worth each of us setting out what our priorities are.

****

SAQ 2.3e	Give two-line definitions of the following computer jargon. Bits Floppy discs ⟶

SAQ 2.3e
(cont.)

Bytes

Winchester discs

Minicomputer

Basic

Real time

Microprocessor

Robotics

Hard copy

RAM

C-language

Microcomputer

Expert systems

User friendly

Binary notation

ROM

Silicon chip

Word processor

Information Technology

Response

Bits – This is the smallest unit of information in a computer such as a switch is on (1) or off (0) which leads to binary notation.

Floppy Discs – These resemble small flexible gramophone records which can receive and deliver large amounts of data via a disc drive. They replaced tapes and cassettes which were slower and less reliable.

Bytes – This is the number of bits the computer uses as its management unit. Early computers were based on 8 bit systems with 16 and 32 bit systems becoming more popular. A byte can define a unit of data such as a number or a piece of text. It is the amount of storage required to store a standard ASCII alphanumeric character. According to the size of the unit of data more than one byte may be needed per unit.

Winchester Discs – A device built into the computer capable of storing in permanent memory the same amount of data as a large number of floppy discs.

Minicomputer – Typically a compromise computer distributed in small numbers through an organisation which are smaller in capacity then a single central main-frame but larger than a greater number of microcomputers. These capacities change but 100, 10 and 1 megabytes are typical orders of magnitude of the three categories.

Basic – A simple, general purpose, computer language available in several 'dialects' with many computers.

Real time – The ability to present information in its final form within a fraction of a second or only a very few seconds of taking the data into the computer.

Microprocessors – A component of a computer or other system incorporating a silicon chip on which a circuit is permanently entered and which can respond to particular electric signal in terms of a predetermined operation.

Robotics – The technology leading to the ability to respond to particular signals from sensors by instructing a machine or robot to carry out actions normally carried out by humans.

Hard Copy – Typically paper containing text, numerical, graphical or diagrammatical information in a form which is more accessible and secure than electronic methods, although less compact and often less retrievable.

RAM – Random Access Memory. This is volatile and wiped clean when the computer is switched off.

C-Language – A high level computer language increasingly used when interfacing computers to spectrometers.

Microcomputer – See minicomputer above.

Expert Systems – A system which contains programs with in-built ability to improve performance with usage by incorporating new information, a form of 'artificial intelligence'.

User Friendly – A self explanatory expression for systems which are recognised as such in comparison with systems found to be 'user hostile'.

Binary Notation – See above for bits. The concept may be difficult to those who believe the decimal system is all important. The younger generation tend to have been taught more facility to operate with any number base other than 10.

ROM – Read only memory. A permanent repository for programs and information in the computer.

Silicon Chip – A tiny, highly pure crystal of silicon, on which one or more large electrical circuits can be permanently printed.

Word Processor – The use of a computer essentially as a type-writer but with many editing and format features for altering the text which may be stored, retrieved or printed. One little recognised disadvantage is that unlike hand-written or typed text the printer does not operate in 'real time'!

Information Technology – A wide ranging term for all forms of computing, radio and telecommunications made possible by recent advances in electronics and software developments towards handling any information with greater efficiency.

One recent definition by the U.K. Department of Trade and Industry is 'The acquisition, processing, storage, dissemination and use of vocal, pictorial, textual and numerical information by a microelectronics-based combination of computing and telecommunications.'

****

SAQ 3.1a

Complete the following table with the wavenumber corresponding to the wavelengths given (all wavelengths are in micrometres, 10^{-6}m). Express the wavenumbers in units of cm^{-1}.

Let me do the first one:

2.50×10^{-6} metre $= 2.50 \times 10^{-4}$ cm, and the reciprocal of this $= 4000$ cm^{-1}, now go ahead and complete the table.

Wavelength (μm)	Wavenumber (cm^{-1})
2.50	4000
3.33	
5.00	
10.00	
15.00	
20.00	
25.00	

Now draw a graph of wavenumber against wavelength in the form below, I have plotted the first and third point for you, is your graph linear?

$\longrightarrow$

**SAQ 3.1a
(cont.)**

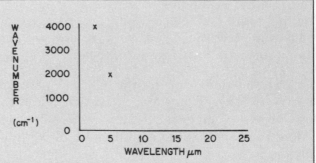

Now find a piece of paper and draw up a linear grid with the scale 4000–400 cm^{-1} along the
abscissa, and write % Transmittance up the ordinate.

This used to be the normal way of presenting an
infrared spectrum. Remember that few ir bands
normally appear in the region 4000–1800 cm^{-1},
with many bands between 1800 and 400 cm^{-1}; do
you see any problems with this type of spectrum
presentation? How would you solve it?

Response

Here is my completed table.

I hope you remembered that there are 100 cm in a metre!

If your numbers vary by a factor of ten, check your units.

Wavelength (μm)	Wavenumber (cm^{-1})
2.50	4000
3.33	3003
5.00	2000
10.00	1000
15.00	666.7
20.00	500
25.00	400

OK, let's plot these as suggested,

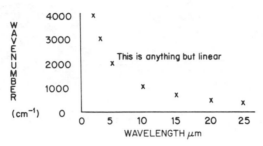

An infrared chart with the abscissa linear in wavenumber would look like:

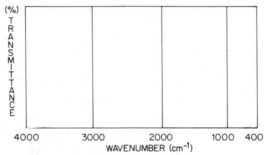

The problem with this is that the interesting regions are below 1800 cm^{-1}, for normal interpretive infrared and as you can see this is crushed up into about a third of the chart.

****

SAQ 3.2a

(*i*)   Why do we not use sodium chloride plates for aqueous solutions?

(*ii*) Why must we watch the pH when using calcium or barium fluoride?

(*iii*) What is a suitable pH range for calcium and barium fluoride windows?

Response

(*i*)   I would not like your budget for infrared plates, they would dissolve!

(*ii*) Acids have a nasty habit of producing HF with inorganic fluorides, not the most healthy environment for infrared studies!

(*iii*) You would have to keep the pH above seven and preferably higher.

****

SAQ 3.2b   Examine the spectra of four solvents, Fig. 3.2b–e (0.02 mm path-length).

Is it possible to use one of these solvents only and get a full-range spectrum without interfering peaks?

Is it possible to use two of these solvents separately and combine the information in the transparent regions? If so, which two solvents?

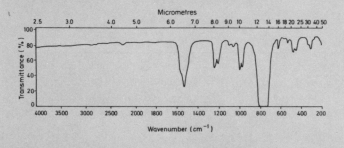

Fig. 3.2b. *Tetrachloromethane*

⟶

SAQ 3.2b
(cont.)

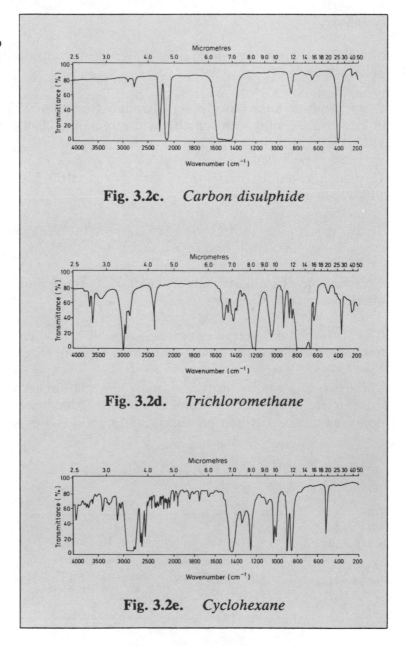

Fig. 3.2c. *Carbon disulphide*

Fig. 3.2d. *Trichloromethane*

Fig. 3.2e. *Cyclohexane*

Response

It is obviously not possible to obtain a spectrum without interfering peaks.

Tetrachloromethane absorbs very strongly (less than 25% transmittance for a 0.2 mm path length) in the regions 720–870 cm^{-1} and 1520–1580 cm^{-1}.

Carbon disulphide absorbs very strongly at 2100–2200 cm^{-1} and 1400–1620 cm^{-1}.

Trichloromethane absorbs very strongly in the regions 2950–3100 cm^{-1}, 1190–1300 cm^{-1} and 650–820 cm^{-1}.

Cyclohexane absorbs strongly at 2650–3000 cm^{-1} plus at least another five narrower regions.

It is possible, however, to cover the range 4000–650 cm^{-1} using tetrachloromethane and carbon disulphide separately, *check it*.

These two solvents are most commonly used. Other commonly used solvents are hexane, trichloromethane and dichloromethane. Solution spectra may be used for solid, liquid and gaseous samples.

****

SAQ 3.2c	Write in the space below two possible combinations of mulling agents, which when used separately allow you to examine the infrared region between 4000 and 600 cm^{-1}.

Response

A combination of iquid paraffin with either of the other two will allow a complete spectrum of a sample to be examined.

****

SAQ 3.3a

Let's suppose you are running an infrared service and three samples await your attention. You can delegate two of the samples below to other workers. Since it is Monday morning and you are not quite awake yet, which would you choose to record? (All spectra are to be recorded using KBr discs).

(*i*) a polycyclic hydrocarbon melting at 125 °C,

(*ii*) phenol,

(*iii*) trimemthylamine hydrochloride.

Response

You should of course take the most difficult sample yourself and set some sort of example!

This is almost certainly phenol. It is a low melting hygroscopic solid. It is bound to be wet if you did not think ahead and desiccate it over the weekend. The chances of getting a good disc are almost nil. I would place a few crystals between two plates and warm the plates, then record the sample as a *melt*.

The amine salt would be difficult, since it would react with the KBr. So if you want the easy one, go for the high melting hydrocarbon. I see no difficulty here!

****

SAQ 3.4a   Which type of solution cell (ie variable path-
                length, demountable or permanent) would you
                consider to be the easiest to maintain?

Response

I would go for the demountable cell every time. These can be easily
dismantled and cleaned. The windows can be repolished, a new
spacer supplied and the cell reassembled.

The permanent cells are difficult to clean and soon become damaged
by water. Path-lengths have to be recalibrated regularly if accurate
work is being undertaken.

Variable path-length cells suffer similar disadvantages and they are
difficult to take apart. The calibration therefore suffers and the cells
have to be calibrated regularly.

****

SAQ 3.4b   Using the interference pattern below, calculate
                the path-length of the cell.

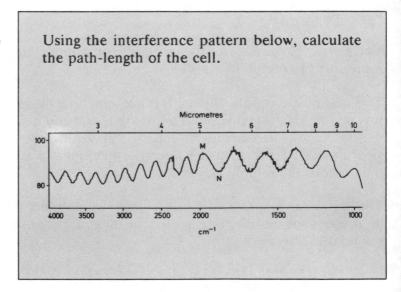

Response

We have to count the number of peak to peak fringes. I chose the region 3780–1180 cm^{-1} since both these frequencies correspond to the tops of peaks.

In this range there are 13 fringes, hence the path-length of the cell, L is calculated as follows:

$$L = \frac{n}{2(\bar{\nu}_1 - \bar{\nu}_2)} = \frac{13}{2 \times 2600} = 2.5 \times 10^{-3} \text{ cm}$$

****

SAQ 3.5a

Why do you think low temperatures lead to sharper bands in an infrared spectrum?

(*i*) Because the concentration of the sample is very low.

(*ii*) Because there is no solute–solute interaction present.

(*iii*) Because there is no rotation at these temperatures.

Response

The critical point is the absence of transitions to excited rotational states, hence the bands are no longer envelopes but sharp single transitions, so answer (*iii*) is correct.

****

SAQ 4.1a	How many vibrational degrees of freedom do the following molecules possess?
	(*i*) Methane (CH_4)
	(*ii*) Ethyne ($CH\equiv CH$)

Response

(*i*) Decide whether the molecule is linear or non-linear, count the atoms and apply the formulae – for linear molecules there are $3N - 5$ and for non-linear molecules $3N - 6$. Methane has 5 atoms and is non-linear and hence has 9 vibrational degrees of freedom.

(*ii*) Here we have four atoms and a linear molecule, hence $(3 \times 4) - 5 = 7$ degrees of freedom.

****

SAQ 4.1b	(*i*) Would you expect the bonds C—N, C=N and C≡N to vibrate at different frequencies when stretched and released?
	If so which would vibrate with the (*a*) highest and (*b*) lowest frequency?
	(*ii*) Which bond would vibrate (stretch and compress) at the highest frequency – a carbon–hydrogen or carbon–chlorine single bond?

Response

Stronger springs vibrate with a higher natural frequency than weak ones – try it!

In molecules the frequency is found to be proportional to the square root of a constant – the force constant – which is related to bond strength.

So the order is $C\equiv N > C=N > C-N$. What about the second part of the question?

Well – Cl is more massive than a hydrogen atom and therefore $C-H$ vibrates with a higher frequency than $C-Cl$. (Big Ben emits a lower note (vibrates more slowly) than the other less massive bells in the Westminster chime collection). You have to be a little wary of this simple argument, we have assumed similar force constants for the two bonds $C-H$ and $C-Cl$.

****

SAQ 4.1c | Given that the $C-H$ stretch vibration for chloroform occurs at 3000 cm^{-1}, calculate the $C-D$ stretch frequency for deuterochloroform.

Response

Looking at the equation:

$$\bar{\nu} = \frac{1}{2\pi c}\sqrt{\frac{f}{\mu}}$$

and assuming that f is the same for both bonds, then we need only calculate the ratio of the reduced masses.

(*i*) for C—H we have

$$\mu = \frac{m_1 m_2}{(m_1 + m_2)} = \frac{12 \times 1}{(12 + 1)} = 0.92$$

(*ii*) for C—D we have

$$\frac{m_1 m_2}{(m_1 + m_2)} = \frac{12 \times 2}{(12 + 2)} = 1.71$$

The vibration frequency is *proportional* to wavenumber and inversely proportional to the square root of these values, hence,

$$\frac{\bar{\nu}_D}{\bar{\nu}_C} = \frac{\sqrt{0.92}}{\sqrt{1.71}} = 0.73$$

So if C—H str for $CHCl_3$ is at 3000 cm^{-1}, C—D str would be expected at:

3000 × 0.73 = 2190 cm^{-1}

∗∗∗∗∗∗∗∗∗∗∗∗∗∗∗∗∗∗∗∗∗∗∗∗∗∗∗∗∗∗∗∗∗∗∗∗∗

SAQ 4.1d	Assuming that the force constants for C≡C, C=C and C—C are in the ratio 3:2:1, and that the normal range for the C=C stretch absorption is 1630–1690 cm^{-1}, what range would you expect for the C—C stretch and C≡C stretch absorptions?

Response

This is quick and easy if you spotted the short-cuts. Don't get depressed if you didn't, but work through the explanation given below. Your understanding of stretching band vibrations will then be pretty good.

$$\bar{\nu} = \frac{1}{2\pi c} \sqrt{\frac{f}{\mu}}$$

and the key to the problem is to recognise that not only are we dealing with ratios, but that the reduced mass is the same for all three cases. We are, of course, dealing with carbon–carbon bonds throughout. Only the force constants differ from triple to double to single bond.

So we can get rid of all the constants μ, c and π and derive a simple equation to use for our calculations.

$$\frac{\bar{\nu}_1}{\bar{\nu}_2} = \sqrt{\frac{f_1}{f_2}}$$

Let's put some numbers into the algebra. If we let f_2 be the force constant for the double bond and f_1, the force constant for the single bond, then the expression under the square root sign becomes the square root of $\frac{1}{2}$ which equals 0.707.

Giving $\bar{\nu}_c$ the values 1630 and 1690 cm^{-1}, we get the range for single bonded carbons to be 1152 to 1195 cm^{-1}. This agrees very well with experimental observations.

A similar calculation for the triple bond, here the constant will be the square root of 1.5, gives 2005 to 2080 cm^{-1}. Triple bond stretching frequencies are usually around 2100 cm^{-1}.

So this does not seem to be a bad model.

****

SAQ 4.1e   Draw diagrams to represent:

(*i*)   the stretching vibration of the O—H bond in an alcohol, ROH

(*ii*) the bending vibration of the carbon skeleton of propane.

(*iii*) the stretching vibration of the C≡N bond in the nitrile RCN.

Response

(*i*)   Well no difficulty here – R—O- - -H $\longleftrightarrow$

(*ii*) This is much more difficult, there are in fact four possibilities! Did you spot this, well done if you got more than one. Here they are:

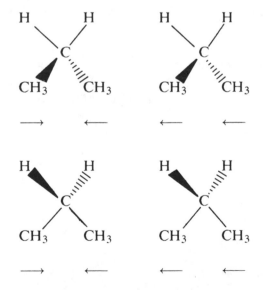

Think about it!

(*iii*) Again no problem here – R≡N

****

SAQ 4.1f

(*i*) Consider the CH_2 fragment of the molecule propane, $CH_3CH_2CH_3$. How many different bending vibrations can occur – 4, 5 or 6?

(*ii*) Now consider water. What is the total number of bending vibrations? Using the →, + and − notation, write them down.

Response

(*i*) There are four as in the example given in the text.

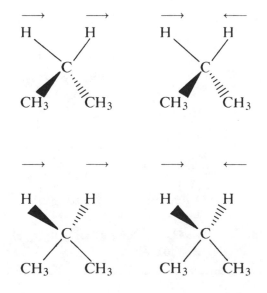

(*ii*) This could be confusing! Did you write four down as follows?

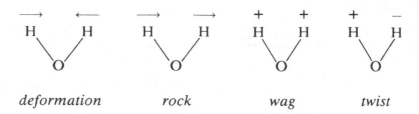

deformation   rock   wag   twist

Only the first of these is a vibration, the others are the three rotations.

So only one is a bending vibration. Remember, if we apply the formula, there are $3N - 6$ vibrations. In this case $(3 \times 3) - 6 = 3$, and two of these are the symmetrical and antisymmetrical stretching vibrations.

****

SAQ 4.2a   Are the following statements true or false?

(*i*)   Given that force constants for bending vibrations are considerably smaller than those for stretching vibrations and that C—O str is typically at 1200 cm^{-1}, C—O bending vibrations are likely to be outside the normal range.

(*ii*)   C—F str occurs around 1400 cm^{-1}, so C—I str is likely to occur at very low frequency.

(*iii*) Vibrations of bonds involving atoms of high atomic number, eg Chromium are likely to be out of range of most routine instruments.

Response

(*i*)   True. Many bending vibrations are outside the normal range. C—H bending vibrations are an exception and can be used for diagnostic purposes in alkenes and aromatic compounds.

(*ii*) True. A quick calculation assuming the force constants to be similar would lead you to the conclusion that C—I stretch would be below 600 cm^{-1}. Since the C—I bond is much weaker than C—F, C—I stretch will be at even lower frequency. So this will also be outside the normal range.

(*iii*) True. This is similar to (*ii*), the more massive the atom, the larger the reduced mass and hence the lower the frequency of absorption.

****

SAQ 4.2b

> In contrast to the carbonyl group, the double bond of ethene $H_2C{=}CH_2$ is electrically symmetrical. Consequently there is *some change/no change* in dipole moment during a stretching vibration, and such a vibration *does/does not* absorb ir radiation and *is/is not* observed in the ir spectrum of the compound.
>
> Write a few lines of explanation, then re-write the sentence above with one of the alternatives.

Response

Ethene has no dipole moment and the excited state species with the C=C bond stretched also has no dipole moment. So we have:

In contrast to the carbonyl group the double bond of ethene

$H_2C=CH_2$ is electrically symmetrical. Consequently there is *no change* in dipole moment during a stretching vibration, and such a vibration *does not* absorb ir radiation and *is not* observed in the ir spectrum of the compound.

****

SAQ 4.2c

Would you expect the $C=C$ str absorption to be *active/inactive*, *weak/strong* in the following compounds.

Place them in order of increasing intensity.

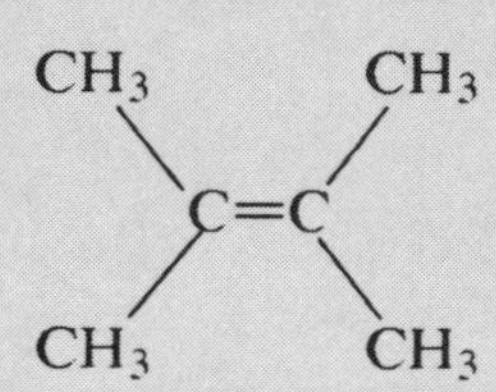

(*a*) 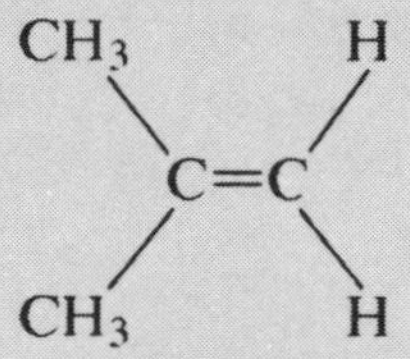   (*b*)

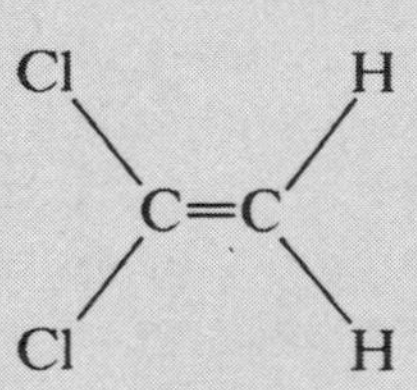

(*c*) 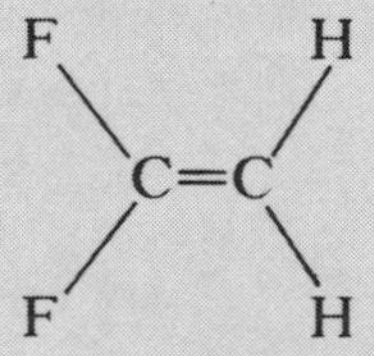   (*d*)

Response

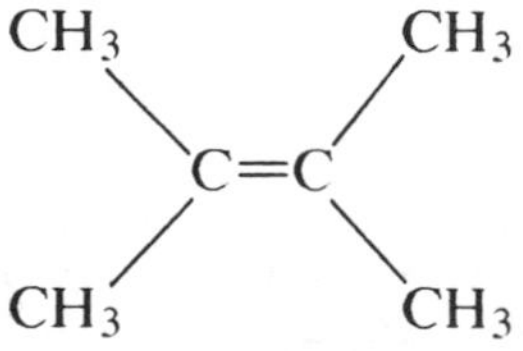

(*a*)   (*b*)
inactive   *weak*

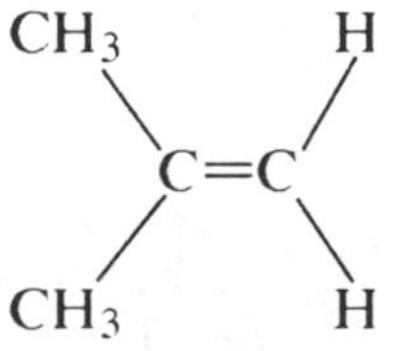

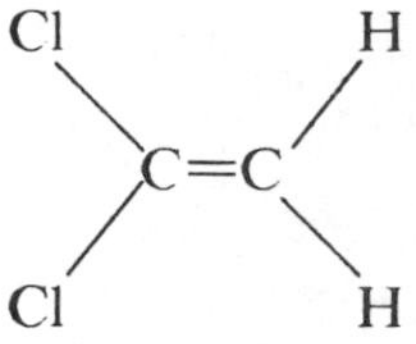

(c)	(d)
stronger	*strongest*

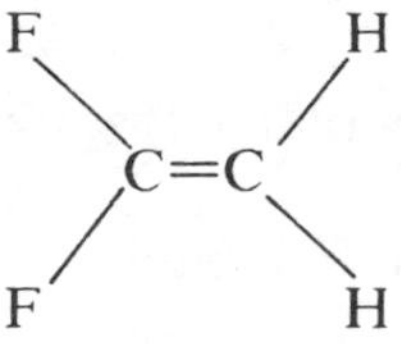

The higher the polarisation of the double bond in the ground state, the larger the change when the bond is stretched. Fluorine has the highest electronegativity, hence (*d*) will have the largest dipole moment. Chlorine will be next then the weakly inductive methyls. So the order was the one given in the question.

****

SAQ 4.2d   Are the following bending vibrations active or inactive?

(*i*)

(*ii*) The twisting vibration in H_2S

(*iii*) H—C≡C—H

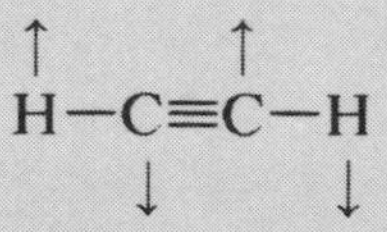

(*iv*) H—C≡C—H

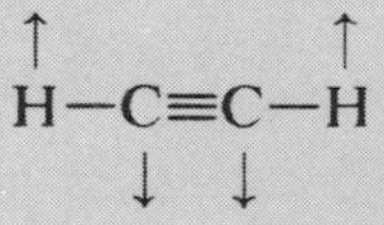

Response

(*i*)   Active, exactly analogous to CO_2

(*ii*)   I meant the following:

How is your vector analysis theory?

The vector in the plane will vary sinusoidally (I think). So there is a change of dipole moment and the mode will be active.

(*iii*) The shape of the molecule after the bending vibration will be:

   there was no dipole before this change, the dipoles still balance, so this is an inactive vibration

(*iv*) Here the shape after the bending is:

   here there is a dipole, where there was none before, so this is active

****

SAQ 4.2e   Would you expect the spectrum of polystyrene to be similar to the spectrum of:

(*i*)   isopropylbenzene : $PhCH(CH_3)_2$

(*ii*) styrene   : $PhCH=CH_2$

(*iii*) 1,3-diphenylethane : $PhCH_2CH_2CH_2Ph$

Response

Polystyrene has the structure:

$$(-CH-CH_2-CH-CH_2-CH-CH_2-CH-CH_2-CH-CH_2-CH-CH_2-)n$$

with Ph substituents and marked * positions and { } brackets:

```
        *           *              {          }
(—CH—CH₂—CH—CH₂—CH—CH₂—CH—CH₂—CH—CH₂—CH—CH₂—)n
    |           |         |         |         |         |
    Ph          Ph        Ph        Ph        Ph        Ph
```

I would expect this compound to have a very similar spectrum to both (*i*) and (*iii*). * to * is similar to (*i*) and the part between { and } is similar to (*iii*).

Styrene will have a C=C str which will be absent in the other spectra.

****

SAQ 4.2f	Consider the vibrational properties of *trans*-2,3-diiodo-2-butene

$$\begin{array}{ccc} CH_3 & & I \\ & C{=}C & \\ I & & CH_3 \end{array}$$

Illustrate each of the three factors that can lead to simplification of infrared spectra.

Response

The bands of one methyl will coalesce with the other.

The C—I str and C—I bending vibrations will be out of range as will all twisting vibrations.

All vibrations symmetrical about the C=C will be inactive, eg C=C str.

****

SAQ 4.3a | A molecule has strong fundamental bands at the following frequencies:

C—H bend at 730 cm^{-1}

C—C str at 1400 cm^{-1}

C—H str at 2950 cm^{-1}

Write down the frequencies (in wavenumbers) of the possible combination bands and the first overtones.

Response

The first overtone bands will occur at double the wavenumber of the fundamentals, so we would expect bands at:

1460, 1800 and 5900 cm^{-1}

The possible combinations are at:

$(730 + 1400) = 2130$ cm^{-1}

$(730 + 2950) = 3680$ cm^{-1}

$(1400 + 2950) = 4350$ cm^{-1}

****

SAQ 4.3b   In the table below I list the bands in the spectrum of SO_2. From the intensities and frequencies can you spot the fundamentals, overtones and combination bands?

Frequency (cm^{-1})	Intensity
519	very strong
1151	very strong
1361	strong
1871	very weak
2305	very weak
2499	moderate

Response

The first three are fundamentals.

$519 + 1361 = 1880$ – so 1871 is a combination band.

$1151 + 1361 = 2512$ – so 2499 is also a combination band.

$2 \times 1151 = 2302$ – so 2305 is an overtone.

****

SAQ 4.3c | Tetrachloromethane is expected to show only four infrared active fundamentals. Three fundamentals absorb at 217 (infrared), 313 (infrared and Raman) and 459 cm^{-1} (Raman only). The fourth is expected to occur in the region 700–800 cm^{-1}. The spectrum has two bands in this frequency range, at 762 and 791 cm^{-1}. Can you account for this observation?

Response

A combination band is possible at $459 + 313 = 772$ cm^{-1}. The fourth fundamental could undergo Fermi resonance with this. It certainly explains the two bands, almost symmetrical about this frequency.

****

SAQ 4.3d | Label the following statements as referring to overtone, combination or Fermi resonance bands.

(*i*) 'Carbonyl stretching bands sometimes are doublets separated by as much as 50 cm^{-1}.'

(*ii*) 'The series of bands in benzene derivatives between 1800 and 2000 cm^{-1} are dependent on the substitution pattern in the ring.'

(*iii*) 'The precise position of this band depends on the frequency of two other bands at lower frequency.'

Response

(*i*)   This may be Fermi resonance. It may also be conformational
        equilibria, hydrogen bonding or coupling in dicarbonyl com-
        pounds.

(*ii*) C—H out-of-plane bending vibrations absorb at about half this
        frequency, also many C—C str frequencies absorb at frequen-
        cies which could combine to give this sort of frequency. These
        are overtones or combination bands.

(*iii*) This must be a combination band.

****

SAQ 4.4a   What is the relationship between vibrational fre-
                quency of a bond and the reduced mass (μ) of
                the two atoms?

                Is the frequency:

                (*i*)   proportional to μ

                (*ii*)   proportional to $\sqrt{\mu}$

                (*iii*) proportional to $\sqrt{\frac{1}{\mu}}$?

Response

Here is the original equation:

$$\bar{\nu} = \frac{1}{2\pi c} \sqrt{\frac{f}{\mu}}$$

and (*iii*) is obviously correct.

****

SAQ 4.4b   Select from the list below the vibration of butane $(CH_3CH_2CH_2CH_3)$ which you would expect to couple strongly with the C—C stretching modes. Remember that there are three factors which determine whether vibrational coupling will take place.

(*i*)   C—H stretch

(*ii*) CH_3 twist

(*iii*) CH_2 rock

(*iv*) CH_2 wag

(*v*)   none of these

Response

Recall the criteria for coupling,

— the groups must be adjacent,

— the frequencies must be similar,

— the vibrations must be in the same plane.

CH_2 rock is in the wrong direction, C—H stretch is at the wrong frequency, as is the twist. CH_2 wag is the only one which meets all criteria. Draw a diagram to convince yourself of this.

****

SAQ 5.2a	Which of the following molecules would you expect to be hydrogen bonded?
	
	Write down a structure for the associated aggregate.
	
	(*i*) SiH_4
	
	(*ii*) Acetic acid (Ethanoic acid)
	
	(*iii*) $CH_3C=O$
	            $\vert$
	            NH_2
	
	(*iv*) Phenol

Response

(*i*)   Silane is analogous to methane, it has no non-bonding electrons hence no hydrogen bonding.

(*ii*) There are lone pairs on the carbonyl oxygen and an electron deficient hydrogen on the O—H group. We therefore have an ideal situation for hydrogen bonding. It is found that carboxylic acids form stable dimers, even in quite dilute solution. The structure of the dimer can be written:

$$CH_3C \underset{O-H\cdots O}{\overset{O\cdots H-O}{<}} CCH_3$$

(*iii*) This compound is solid at room temperature, which is very unusual for so low a relative molar mass. There is a lone pair present on the carbonyl oxygen and an electron deficient hydrogen on nitrogen, an ideal situation.

The structure of the dimer can be written in a least two ways:

$$
\begin{array}{cc}
 & CH_3 \\
 & | \\
CH_3 & C=O \\
| & | \\
C=O\cdots H-NH \\
| \\
NH_2
\end{array}
\qquad
CH_3C \underset{N-H\cdots O}{\overset{O\cdots H-N}{<}} \begin{array}{c} H \\ CCH_3 \\ H \end{array}
$$

The second of these is analogous to the structure I wrote above for ethanoic acid.

If you look at the first of these structures, it is possible to add another molecule of ethanamide to form a trimer:

$$
\begin{array}{ccc}
 & & CH_3 \\
 & CH_3 & | \\
 & | & C=O \\
CH_3 & C=O\cdots H-NH \\
| & | \\
C=O\cdots H-NH \\
| \\
NH_2
\end{array}
$$

This can be extended to form tetramers and higher aggregates. This is also possible with ethanoic acid.

You may have learnt elsewhere that hydrogen bonding is critical in the tertiary structure of proteins. Proteins are polyamides, and interactions between protein chains analogous to the intermolecular interactions I described above are responsible in large part for the three dimensional shape of these large molecules.

(*iv*) Phenol is a low melting solid. It has a higher melting point than compounds of similar molar mass eg toluene (methylbenzene). There is likely to be molecular association present again. We can write the structure of the dimer as follows:

Again further association to trimers etc. is possible.

****

SAQ 5.2b | Classify the following compounds as *inter*molecularly hydrogen bonded, *inter* and *intra*molecularly hydrogen bonded or lacking hydrogen bonds.

(*i*) ethanol,

(*ii*) 2-aminobenzoic acid,

(*iii*) chloroform,

(*iv*) ethyl acetoacetate ($Ch_3CCH_2COCH_2CH_3$).

$$\underset{\|}{O} \qquad \underset{\|}{O}$$

Response

(*i*) Is intermolecularly hydrogen bonded – there is no possibility of any intramolecular bonding.

(*ii*) Is set up for both inter and intra. I draw below two possible intramolecular hydrogen bonds. The first is much more likely than the second which would involve loss of conjugation of the nitrogen lone pair with the benzene ring because of rotation.

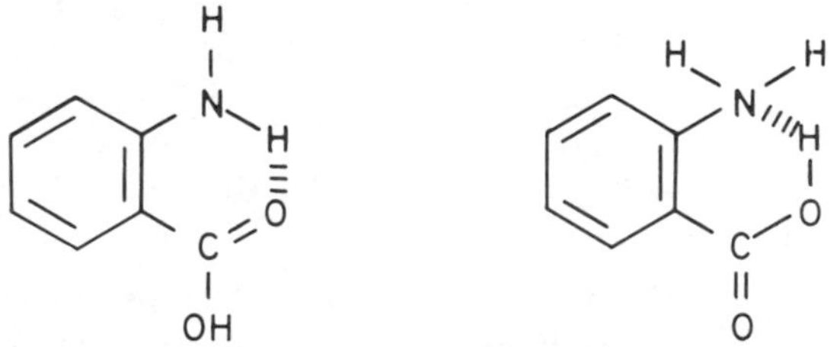

(*iii*) Has a donor hydrogen but no suitable lone pair exists in the molecule to form a strong hydrogen bond. A very weak hydrogen bond could form with the chlorine lone pairs. So this is a marginal case of intermolecular hydrogen bonding.

(*iv*) In the keto-form no hydrogen bonding is possible. You may remember that this molecule exhibits keto-enol tautomerism. The enol form is intramolecularly hydrogen bonded. The presence of this bond is often used as an explanation of the stability of the enol form. Note the six-membered ring.

****

SAQ 5.3a   (*i*) Would you expect the O—H stretching mode to vary in intensity when hydrogen bonded? (You will need some of the ideas from Part Four.)

(*ii*) If so would it increase or decrease?

Response

The intensity of an infrared absorption depends on the change in dipole moment during the vibration. When a hydrogen atom is held in a hydrogen bond it is in a much more polar environment than when unchelated. The hydrogen atom is moving between two polar centres,

$$\longleftrightarrow$$

$$O-H \ldots O$$

I think it is reasonable to expect an increase in intensity from dimeric and higher polymeric species.

****

SAQ 5.4a	Examine the ir spectra of hexan-1-ol given in Fig. 5.4c and Fig. 5.4d. (*i*) Assign the monomer and polymer bands in the O—H stretching region. (*ii*) How many bands are there in this region for each concentration? (*iii*) Note any differences from the ethanol spectra you have just studied and comment on them.

Response

(*i*) The monomer band is at 3640 cm^{-1} and the polymer band at 3000–3600 cm^{-1}.

(*ii*) There are two bands in the 10% solution and three in the 1% solution.

(*iii*) The three bands are much more obvious in these spectra compared to the ethanol ones. The O—H stretching bands are weaker. We will return to this example in the Quantitative

analysis Part of the unit and see that we are making dangerous
assumptions.

****

SAQ 5.4b   Examine the spectra of hexanoic acid given in
Fig. 5.4e and Fig. 5.4f

(*i*)   Assign the monomer and polymer bands in
the O—H stretching region.

(*ii*) Assign the monomer and polymer bands in
the C=O stretching region.

(*iii*) Compare the shape of the O—H stretching
bands with those for ethanol, do you have
any explanation for the differences?

Response

(*i*)   The monomer band is at 3540 cm^{-1} and the polymer band at
2300–3500 cm^{-1}.

(*ii*) The monomer band is at 1760 cm^{-1} and the polymer band at
1720 cm^{-1}.

(*iii*) The O—H str band in acids is always broader than in alcohols. There has been extensive discussion in the literature on
the reasons for this. Explanations have included the superim-
position of the O—H stretching frequencies of many different
species, Fermi resonance and coupling with other modes.

****

SAQ 5.5a   Is it possible to use the presence of an intramolecular hydrogen bond to distinguish the
following pairs of isomers?

Write a structure for the intramolecularly hydrogen bonded isomer.

(*i*)   2-nitrophenol from 4-nitrophenol.

(*ii*) trans-1,4-dihydroxycyclohexane from cis-
1,4-dihydroxycyclohexane (you may need
to build models of these compounds).

(*iii*) the isomers of the alkaloid, granatinol:

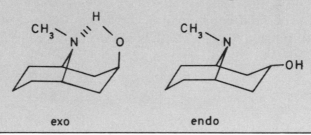

exo   endo

Response

(*i*)   2-nitrophenol forms a strong intramolecular hydrogen bond,
as shown below:

The groups in the 4- isomer are wrongly arranged spatially for intramolecular bond formation. Study of low concentration solutions of these two isomers would therefore distinguish them. Incidentally, a shift of the two stretching frequencies (sym and anti-sym str) of the NO_2 group is also observed. In the 2- isomer these bands are at 1550 and 1315 cm^{-1}, while in the 3- isomer they absorb at 1567 and 1350 cm^{-1}. Note the usual shift to lower frequency in the hydrogen bonded isomer.

(*ii*) The trans 1,4- isomer can be in the diaxial or diequatorial conformer, in both of which it is sterically impossible for the groups to form an intramolecular hydrogen bond. The cis isomer can only be axial-equatorial. So neither isomer can form intramolecular hydrogen bonds *in the chair conformation.*

In the boat conformation the trans isomer can however form an intramolecular bond as shown in the diagram below.

(*iii*) Again here we have to write the isomers in the chair/boat conformation. The exo-isomer can then form an intramolecular hydrogen bond while the endo-isomer cannot. This technique was used in the structure determination of these isomers.

exo   endo

This part of the question was difficult and relied on your prior knowledge of organic chemistry, so don't worry if you did not see this, but note the power of the technique.

****

SAQ 6.1a	If the units used for concentration and path-length are mol dm^{-3} and mm respectively, what are the units of molar absorptivity?

Response

Beer's Law states that Absorbance $= \epsilon cl$

therefore $\epsilon = \dfrac{A}{cl}$

Absorbance is dimensionless, c, the concentration is in mol dm^{-3}; and the path-length is in mm, so the units are

$$\frac{1}{\text{mol dm}^{-3} \times \text{mm}}$$

To convert mol dm^{-3} to mol m^{-3}, we must multiply by 1000, and similarly to convert mm to m we must multiply by 1000. So the corrections cancel and the units are,

$$\frac{1}{mol\ m^{-3} \times m} = \frac{1}{mol\ m^{-2}} = m^2\ mol^{-1}$$

NOTE This is the recognised SI unit for molar absorptivity but path-lengths of one or two cm are often used, especially in the visible and ultraviolet regions. So when comparing absorptivity values look at the units carefully; $m^2\ mol^{-1}$ is not always used.

****

SAQ 6.1b

> Try to write down the Beer–Lambert Law in a form that relates % transmittance to concentration and path-length. Is the constant of proportionality in this equation the same as in the equation involving absorbance?

Response

This is a mathematical problem but is worth asking since many infrared instruments operate only in transmittance, so both equations are needed.

The Beer–Lambert Law is expressed as

$A = \epsilon cl$

We now need an expression linking absorbance to transmittance and the two relevant equations are

$T = -I/I_0$ and $A = \log_{10} I_0/I$

hence $A = \log_{10} 1/T$

therefore $A = \log_{10} 1/T = \epsilon cl$

Instruments are usually calibrated in % T and our equation becomes

$$A = \log_{10} \frac{100}{\%T} = \epsilon c l$$

alternatively

$$A = -\log_{10} \frac{\%T}{100} = \epsilon c l$$

****

SAQ 6.1c	A compound has an absorption band at 830 cm^{-1} with a molar absorptivity of 2.0 m^2 mol^{-1}, calculate the range of concentrations that would give a satisfactory analytical peak using a pathlength of 1.0 mm. (I mean a minimum absorbance of 0.2 and a maximum of 0.5.)

Response

The absorbance of a solution $= \epsilon c l$

Here $A = 0.2$ for the lower limit, $l = 1.0$ (in mm, remember) and $\epsilon = 2.0$ m^2 mol^{-1}.

therefore,

$$c = \frac{0.2}{1.0 \times 2.0} = 0.1 \text{ mol dm}^{-3}$$

for the upper limit,

$A = 0.5, l = 1.0$ and $\epsilon = 2.5$ hence,

$$c = \frac{0.5}{1.0 \times 2.0} = 0.25 \text{ mol dm}^{-3}$$

Note the high concentrations compared to uv/visible spectrometry if you have any experience of the technique. This is because of the shorter path-lengths used and the low molar absorptivity. The value in this example is in fact quite high for an infrared band. The technique is therefore inherently less sensitive.

****

SAQ 6.1d   A 1.0% w/v solution of hexan-1-ol has an absorbance of 0.37 at 3660 cm^{-1} in a 1.0 mm cell. Calculate its molar absorptivity at this frequency.

Response

The only snag here is the concentration units used. These must be expressed in mol dm^{-3}.

1.0% w/v means 1.0 g made up to 100 cm^3, ie 10 g dm^{-3}. The relative molecular mass of hexan-1-ol ($C_6H_{13}OH$) is 72 + 13 + 16 + 1 = 92, so this is a 10/92 = 0.11 mol dm^{-3}.

Since $A = \epsilon cl$ then $\epsilon = A/cl$

hence $\epsilon = 0.37/(0.11 \times 1.0) = 3.36 \text{ m}^2 \text{ mol}^{-1}$

$\epsilon = 3.4 \text{ m}^2 \text{ mol}^{-1}$.

in keeping with the precision of the other parameters used in the calculation, such as absorbance and concentration.

****

SAQ 6.1e Measure the absorbance of the band marked A on Fig. 6.1c.

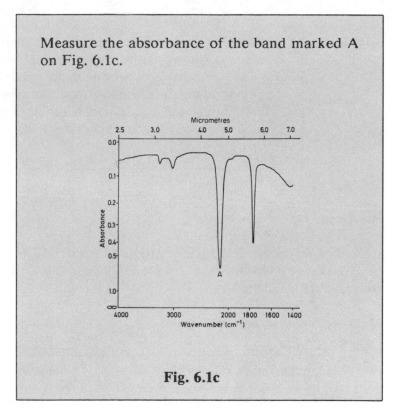

Fig. 6.1c

Response

It is difficult without a scale, but the value I got was 0.584. It is unlikely that you agree exactly with this; if you do then you must have some experience in this field. Let's go back to the main text and I will explain how I arrived at this value.

****

SAQ 6.1f

I would like you to now go through the whole analysis process with real spectra. Fig. 6.1e (*i*)–(*v*) are infrared spectra of phenylacetylene (phenylethyne) in tetrachloromethane solution. Please determine the concentrations of phenylacetylene shown in the spectra in Fig. 6.1f, as % v/v.

Response

First choose a peak. The spectra provided have only been recorded over the range 3400–3000 cm^{-1}. I would choose the C—H stretching frequency at 3320 cm^{-1}.

Measure the absorbances, I got

Concentration (% v/v)	Absorbance
1.0	0.105
2.0	0.230
3.0	0.310
4.0	0.410
5.0	0.520

I have plotted absorbance against concentration in Fig. 6.1i below.

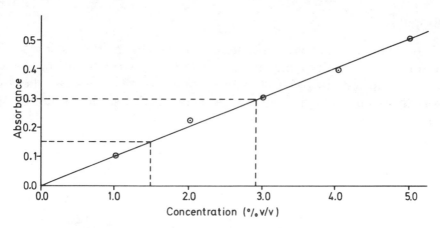

Fig. 6.1i. *Beer's Law plot for the 3320 cm^{-1} band of phenylacetylene*

The unknown solutions have absorbances of 0.29 and 0.16. This gives concentrations read from the graph of 2.84 and 1.49% v/v.

****

SAQ 6.1g

Fig. 6.1g (*i*)–(*v*) are infrared spectra of propan-2-ol as 5, 4, 3 2 and 1% v/v solutions in CCl_4. Draw a Beer–Lambert plot from measurements of the O—H stretching band.

Is this graph linear? Could you use it as a calibration curve for the determination of propan-2-ol?

Response

First we have to choose a frequency, I have chosen the most intense absorption of the polymer band of the O—H stretching frequency. I have marked this on Fig. 6.1g to help you. I then read the absorbances and obtained the results below, which I have plotted in Fig. 6.1j.

Propan-2-ol in CCl_4 (%v/v)	Absorbance
5.0	0.398
4.0	0.275
3.0	0.155
2.07	0.081
1.0	0.020

You can see that this plot is anything but linear. It is also temperature dependent so O—H str is not a good choice as an analytical band.

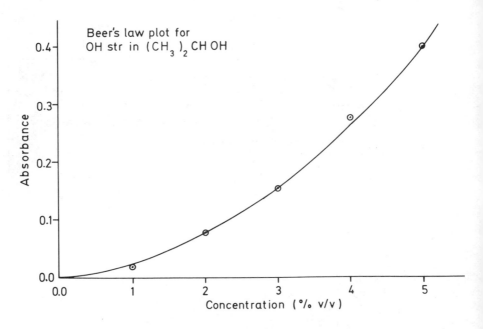

Fig. 6.1j.   *Beer–Lambert Law plot for propan-2-ol in CCl_4*

| SAQ 6.1h | It is possible to successfully use the O—H stretching absorption as an analytical peak. Suggest a way of doing this (your method will have to remove hydrogen bonding effects). |

Response

This was probably very difficult for you, but hopefully made you think about the contents of the previous Part of the Unit. We need *one* peak. In most hydrogen bonded materials we observe one band from the monomer and one or more absorption bands from molecularly associated polymers are observed. In most analytical work the concentration is fairly low. The trick is to use a solvent that forms *strong* hydrogen bonds to the analyte. One of the commonest used is pyridine, which hydrogen bonds to alcohols through the lone pair electrons on the nitrogen. This leads to one band in the O—H stretching region, since all the solute molecules now participate in H-bonding (with the solvent).

N: H—O—R

****

SAQ 6.2a   Draw a plot of absorbance against concentration
from the data below and calculate the molar absorptivity in units of $m^2 \, mol^{-1}$.

Acetone in CCl_4 (%v/v)	% Transmittance at 1719 cm^{-1}
0.25	65.6
0.50	48.5
1.00	26.9
1.50	16.0
2.00	10.0

The transmittance values given were read
straight from the spectrum. The baseline had a
transmittance of 86% at 1719 cm^{-1}.

The density of acetone is 0.790 $g.cm^{-3}$.

The path-length was 0.1 mm.

Warning – this may take you half an hour, so
this may be a good time to break.

Response

That are a few complications here compared to the former example.
You have a few calculations to do before you can draw the graph.

First we must convert % transmittance to absorbance. I hope you
spotted this one otherwise your graph would have been anything but
linear!

I hope you remembered that

Absorbance $= -\log_{10}$ (%Transmittance/100)

or

$$A = -\log_{10} (\%T/100),$$

taking the first result as an example this gives,

$$A = -\log_{10} (65.6/100) = -4\log_{10} (0.656) = 0.183$$

working the others out in the same way gives,

Acetone in CCl_4 (%v/v)	% Transmittance at 1719 cm^{-1}	Absorbance * at 1719 cm^{-1}
0.25	65.6	0.183
0.50	48.5	0.315
1.00	26.9	0.570
1.50	16.0	0.796
2.00	10.0	1.000

You were told that the baseline was at 87% transmittance, this corresponds to an absorbance of 0.06. We now must subtract this from the absorbances in the table, giving the actual absorbance of the peak in each spectrum.

% Acetone in CCl_4 (v/v)	% Transmittance at 1719 cm^{-1}	Absorbance * at 1719 cm^{-1}
0.25	65.6	0.123
0.50	48.5	0.225
1.00	26.9	0.510
1.50	16.0	0.736
2.00	10.0	0.940

* This peak is the C=O stretching band.

We now come to the second problem, the units of concentration in this example are in % v/v. We can therefore convert all the data above to mol dm^{-3} or be lazy and draw the graph in % v/v units and then convert the value of the absorptivity afterwards. In a real application the choice here will depend on the units you would like the concentration of unknown samples to be expressed in. If you require % v/v then you would stay with these units. If you wanted mol dm^{-3} then it is less time consuming to do the calculations now.

I will do it both ways to ease your frustration with this answer!

As an example I will take the first result again, here we have a concentration of 0.25% v/v, ie 0.25 cm^3 in 100 cm^3. This is 2.5 cm^3 in one dm^3. 2.5 cm^3 has a mass of volume × density, that is 2.5 × 0.790 = 1.98 g.

The relative molecular mass of acetone is 58, so this is 1.98/58 = 3.41 × 10^{-2} mol dm^{-3}.

Here is that table again, duly completed for all solutions.

Acetone in CCl$_4$ (%v/v)	Acetone conc. (mol dm^{-3})	% Transmittance at 1719 cm^{-1}	Absorbance at 1719 cm^{-1}
0.25	3.41 × 10^{-2}	65.6	0.123
0.50	6.83 × 10^{-2}	48.5	0.255
1.00	0.137	26.9	0.510
1.50	0.205	16.0	0.736
2.00	0.273	10.0	0.940

The two calibration graphs are given below in Figs. 6.2c and Fig. 6.2d.

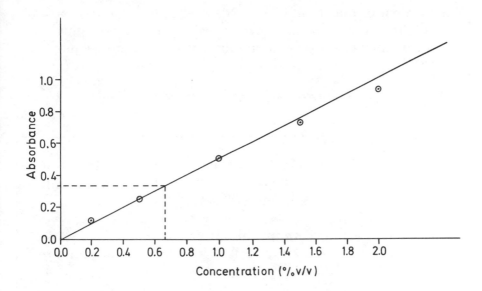

Fig. 6.2c. *Calibration graph for the determination of acetone by infrared spectrometry [C=O band 1719 cm⁻¹, path length 0.1 mm]*

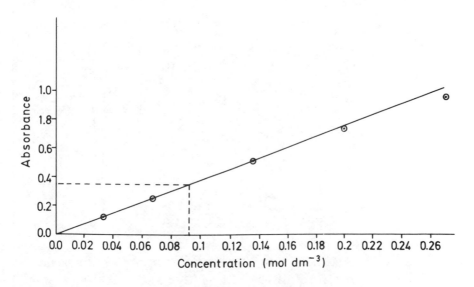

Fig. 6.2d. *Calibration graph for the determination of acetone by infrared spectrometry [C=O band 1719 cm⁻¹, path length 0.1 mm]*

The gradient of the graph (Fig. 6.2d) is 3.75, but this was for a 0.1 mm cell, the units we want assume pathlength in mm, so this value must be divided by 0.1. The molar absorptivity therefore equals 37.5 m^2 mol^{-1}.

****

SAQ 6.2b

The infrared spectrum of a 10% v/v solution of commercial propan-2-ol in CCl_4 in a 0.1 mm path length cell is shown in Fig. 6.2e.

(*i*) Determine the concentration (mol dm^{-3}) of acetone in this solution using the calibration curve plotted in SAQ 6.2a.

(*ii*) perhaps more important, calculate the % acetone in the propan-2-ol.

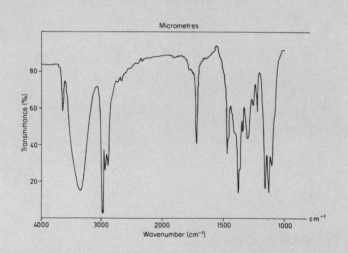

Fig. 6.2e. *Infrared spectrum of commercial propan-2-ol in CCl_4 [10% v/v, 0.1 mm pathlength]*

Response

The spectrum shows a C=O stretching band with a transmittance of 40.5%. The baseline has a transmittance of 88%. This gives a corrected absorbance of $(0.393 - 0.056) = 0.337$.

From the calibration graph this corresponds to 0.67% v/v or 0.09 mol dm^{-3}. This is a 10% v/v solution in CCl$_4$, so we must multiply by ten, giving a concentration of acetone in propan-2-ol of 0.90 mol dm^{-3} or 6.7% v/v. The purity by volume of the propan-2-ol is 93.3%.

****

SAQ 6.3a

Measure the absorbance of the xylenes (*o m* and *p*) at 740, 770 and 800 cm^{-1} in the spectra given in Fig. 6.3a, b and c.

The infrared spectrum of a commercial sample of xylene is given in Fig. 6.3d. Estimate the concentrations of the three isomers in this sample.

All the spectra were recorded using the same cell which had a path-length of 0.1 mm.

Note that the solvent used in all cases was cyclohexane but that the concentrations of the four samples was not always the same:

o-xylene 1% v/v
m-xylene 2% v/v
p-xylene 2% v/v
commercial xylene 5% v/v

There are a few short-cuts that I hope you will spot. If not they are in the answer.

Response

I read the absorbances as:

o $0.440 - 0.012 = 0.428$ Let $E_o = 0.428$

m $0.460 - 0.015 = 0.445$ Let $E_m = 0.445/2 = 0.223$

p $0.545 - 0.015 = 0.530$ Let $E_p = 0.5330/2 = 0.265$

I have halved the values for m- and p- since the solutions are twice as concentrated.

These numbers are proportional to the molar absorptivity, hence the concentrations of the xylenes can be estimated in the mixture, once the absorbances are measured.

These are:

At 740 $cm^{-1} = 0.194 - 0.038 = 0.156$

At 770 $cm^{-1} = 0.720 - 4\ 0.034 = 0.686$

At 800 $cm^{-1} = 0.133 - 4\ 0.030 = 0.103$

Dividing these absorbances by the E values above gives the % v/v of each isomer in the mixture.

We have:

o-xylene in mixture $= 0.156/0.428 = 0.364\%$ v/v

m-xylene in mixture $= 0.686/0.223 = 3.076\%$ v/v

p-xylene in mixture $= 0.103/0.265 = 0.389\%$ v/v

Put another way, for o-xylene, if a 1% v/v solution gives $A = 0.428$, then because $c \propto A$ (ϵ, l constant), then $A = 0.156$ which gives the concentration equal to

$$1 \times \frac{0.156}{0.428} = 0.364\% \text{ v/v.}$$

This is a 5% v/v solution and these three isomers total only 3.829% v/v. This is a very large discrepancy and there are obviously errors present. Let's go back to the main text and discuss the possible sources of this error.

****

SAQ 6.3b   Can you think of three reasons for the discrepancy between the calculated values of the concentration of the isomers and the total concentration?

Response

I have told you that there is a fourth isomer present, so I suppose that could account for the difference.

We have also ignored any non-linearity of the Beer's Law plots, having used one value to get the constant of proportionality above for each solution.

I found it difficult to select a baseline for the analysis because of overlapping peaks. You therefore almost certainly selected a different baseline and this could lead to different answers.

Let's go on to the next stage.

****

SAQ 6.3c	How would you go about finding the constants in the equation above, ie

$$\% trans = K(A_{965}/A_{1163}) - f$$

Response

This is the equation of a straight line of gradient K and y-intercept $-f$ if we plot $\% trans$ against the ratio of absorbances at the two frequencies, 965 and 1163 cm^{-1}. So we have to prepare a series of solutions of known *trans* composition, measure the ratio of absorbances and plot the $\% trans$ content against this.

****

SAQ 7.1a	Examine the spectra in Figs. 7.1a–7.1e and classify them as below;

    (*i*)   aliphatic C—H bonds only,

    (*ii*) aliphatic and aromatic C—H bonds,

    (*iii*) an alkene or aromatic compound containing no aliphatic C—H bonds,

    (*iv*) an alkyne,

    (*v*) a deuterated compound.

Response

These compounds can be classified using the C—H or C—D stretching frequencies.

Fig. 7.1a contains a series of C—H stretching bands both above and below 3000 cm^{-1}. This is therefore the spectrum of a compound containing both aliphatic and aromatic C—H bonds.

Fig. 7.1b has no absorption above 3000 cm^{-1} and is of type (*i*).

Fig. 7.1c has only weak absorption around 3000 cm^{-1} but has C—D stretching bands at 2120 and 2250 cm^{-1}, this is a deuterated compound.

Fig. 7.1d has aromatic C—H stretching bands but also has the strong band at 3300 cm^{-1} indicative of the C—H stretching frequency of an alkyne.

Fig. 7.1e has no aliphatic absorptions below 3000 cm^{-1}, and is either an aromatic compound containing no aliphatic hydrogen or a simple alkene.

****

SAQ 7.1b

> Carbon monoxide absorbs at 2143 cm^{-1}. What does this tell us about the bond order in this molecule?

Response

This is close to the carbon—carbon triple bond absorption frequency, indicating a bond order of three for carbon monoxide.

****

SAQ 7.1c   Examine the three spectra in Figs. 7.1g–7.1i.

They are of 1,2-dimethyl, 1,3-dimethyl and 1,4-dimethylbenzene, in the region 2000–1650 cm^{-1}.

Which is which?

Note that these are absorbance spectra.

Response

Comparison with the 'standard patterns' in Fig. 7.1f gives the following:

Fig. 7.1g.   1,2-dimethyl

Fig. 7.1h.   1,3-dimethyl

Fig. 7.1i.   1,4-dimethyl

****

SAQ 7.1d   Would you expect C—O stretch to be more or less intense than C—C stretch?

Response

The intensity of a band in the infrared region depends on the change in dipole moment during the vibration. The change for C—O will be much greater than for C—C, hence we would expect an intense absorption for C—O stretch.

****

SAQ 7.1e	Absorption in the fingerprint region occurs at variable frequency for C—C and C—O stretching frequencies. Recalling the rules that were put forward in Part Four for coupling between vibrations, would you expect C—H out-of-plane deformations to couple with these stretching vibrations?

Response

The answer must be no, since as I said in the text, these out-of-plane absorptions can be relied on for interpretation purposes. But why don't they couple? The vibrations are out-of-plane and therefore at right angles to skeletal stretching vibrations. One of the 'rules' for coupling is that the extent of coupling decreases as the angle between them increases. Coupling is at a maximum when the angle is zero and drops to nothing if the bonds are at right angles.

****

SAQ 7.2a	Which of the following compounds contain a carbonyl group? Look up their structures if you need to. (*i*) benzoic acid (*ii*) ethanal ⟶

SAQ 7.2a
(cont.)

(*iii*) ethanol

(*iv*) urea

(*v*) pentan-2-one

(*vi*) ethanoyl chloride

(*vii*) acetic anhydride

(*viii*)picric acid

(*ix*) ethyl acetoacetate

Response

They all contain a carbonyl group except ethanol and the badly named picric acid. Ethanol is an alcohol and contains the O—H group, picric acid is a nitro phenol, the nitro group increases the acidity of the phenolic O—H group, hence the name.

(*i*)   Benzoic acid contains the carboxyl group:

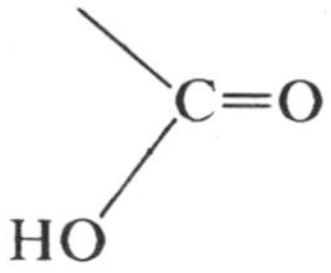

$$\begin{array}{c} \diagdown \\ \diagup \end{array}\text{C=O}$$
$$\text{HO}$$

(*ii*)   Ethanal, perhaps better known by its trivial name, acetalde-hyde is an aldehyde and contains the aldehyde group:

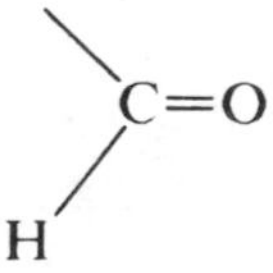

$$\begin{array}{c} \diagdown \\ \diagup \end{array}\text{C=O}$$
$$\text{H}$$

(*iv*) Urea is an amide and contains the amide group: C=O

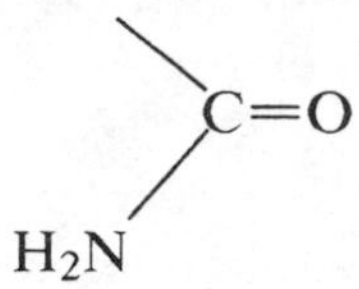

(*v*) Pentan-2-one is a ketone and contains the keto group:

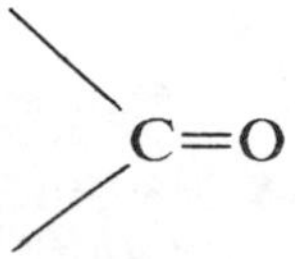

(*vi*) Ethanoyl chloride, perhaps better known as acetyl chloride is an acid chloride and contains a carbonyl group next to a chlorine atom:

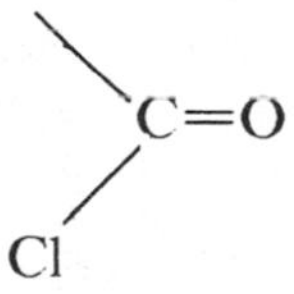

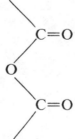

(*vii*) Acetic anhydride contains the group:

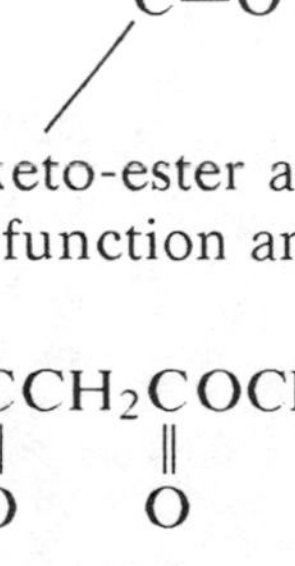

(*ix*) Ethyl acetoacetate is a keto-ester and contains two carbonyl groups, one in the ester function and one a keto group:

$$CH_3\underset{\underset{O}{\|}}{C}CH_2\underset{\underset{O}{\|}}{C}OCH_2CH_3$$

****

SAQ 7.2b

The following are examples of an ester, amide, ketone, aldehyde, acid chloride and acid anhydrides

CH_3COOEt CH_3CONH_2 CH_3COCH_3

CH_3CHO CH_3COCl $CH_3CO-O-COCH_3$

With reference to these examples and to the correlation table in Fig. 7.2a, place the functional group classes in order of increasing frequency for the C=O stretch.

I will start you off: amide <

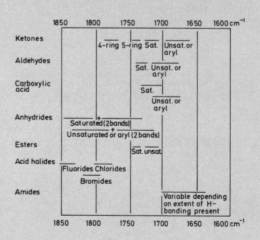

Fig. 7.2a.   *Carbonyl stretching frequencies*

Response

These are all saturated non-conjugated compounds.

From the correlation table for carbonyl frequencies it should be obvious that the order is,

amides < ketones < aldehydes < esters < acid chlorides < acid anhydrides.

****

SAQ 7.2c

Are the following statements true or false?

(*i*) Double bonds are shorter than triple bonds and have larger force constants.

(*ii*) Triple bonds absorb in the infrared region at shorter wavelengths than single bonds since they have larger force constants.

(*iii*) the group X lowers the C—R bond order in X—C—R, it will raise its stretching frequency.

Response

(*i*) The length of bonds follows the order single < double < triple. So double bonds are *longer* than triple bonds, they will therefore have smaller force constants. False on two counts.

(*ii*) Triple bonds do have larger force constants, and this means that they will absorb at higher frequency, ie shorter wavelengths. True.

(*iii*) If X lowers the bond order, the C—R bond will lengthen thus the stretching frequency will decrease. False.

****

SAQ 7.2d

You found in SAQ 7.2b that the carbonyl stretching frequency increased in the order amides, ketones, aldehydes, esters, acid chlorides. Use this information to deduce whether the inductive or resonance effect is the dominant effect

where X =O, N, Cl in the molecule

$$\begin{array}{c} R \\ \diagdown \\ \diagup C = O \\ X \end{array}$$

by comparison with the case where X = C (ketones).

Response

You should have found that the resonance effect was dominant where X = N (amides). In the case of acid chlorides (X = Cl) and esters (X = O), the inductive effect is dominant. This is so because amides absorb at frequencies less than ketones, while esters and acid chlorides absorb at higher frequencies.

****

SAQ 7.2e | Place the following compounds in order of *increasing* carbonyl stretching frequency.

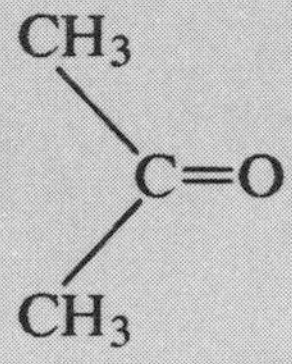

$$CH_3 \diagdown C=O \diagup CH_3$$

(i)

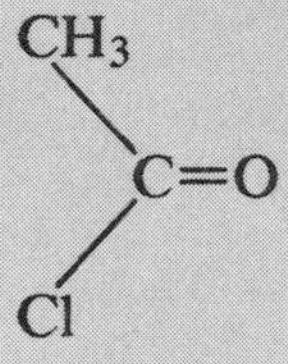

$$CH_3 \diagdown C=O \diagup Cl$$

(ii)

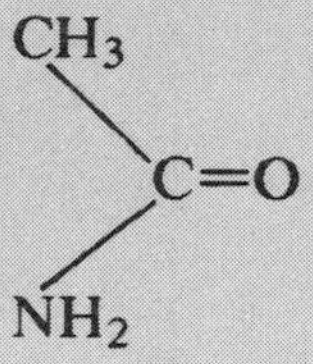

$$CH_3 \diagdown C=O \diagup NH_2$$

(iii)

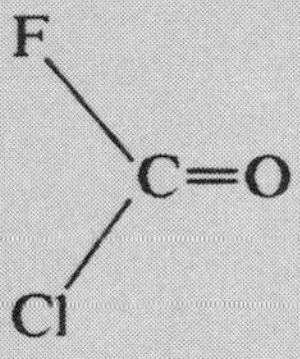

$$F \diagdown C=O \diagup Cl$$

(iv)

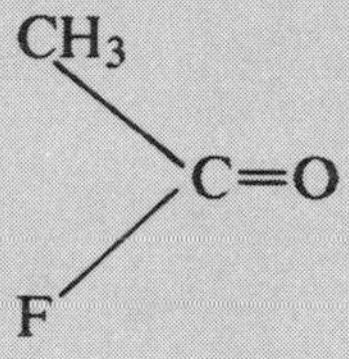

$$CH_3 \diagdown C=O \diagup F$$

(v)

Response

The lowest will be acetamide (*iii*). You saw in the text that this was the only group which showed a dominant mesomeric effect.

All others will absorb at higher frequency than the ketone (*i*).

The inductive effect now dominates and the effect of a chlorine atom is less than a fluorine atom therefore (*ii*) will absorb at lower frequency than (*v*), while (*ii*) will show the greatest effect and absorb at the highest frequency.

The order is therefore, (*iii*), (*i*), (*ii*), (*v*), (*iv*).

****

SAQ 7.2f   Arrange the following molecules in order of increasing carbonyl stretching frequency,

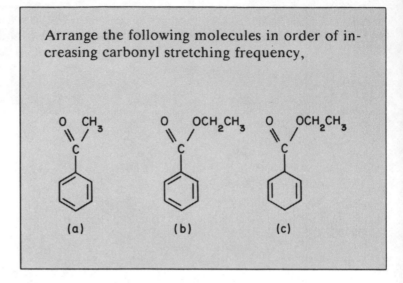

(a)   (b)   (c)

Response

The order you should have arrived at is $(a) < (b) < (c)$.

(b) is conjugated and will be at lower frequency than its non-conjugated analogue (c). (a) is a ketone and will absorb at lower frequency than the ester (b).

****

SAQ 7.2g	Bearing in mind the typical carbonyl stretching frequencies of cyclic ketones for 3, 4, 5 and 6 membered ring systems (1815, 1775, 1750 and 1718 cm^{-1} respectively), can you make an educated guess at the $C=O$ stretching frequency in the compound below?

Response

The frequency is found to be 1785 cm^{-1}. This is consistent with the rigid structure of the molecule and indicates a ring size of 3–4. This is reasonable since there is likely to be a little more 'give' in this system than in cyclopropanone.

****

SAQ 7.2h   Assign the appropriate C=O stretching frequencies to the structures below, from the following
values:

1775,   1750,   1745,   1700   cm^{-1}

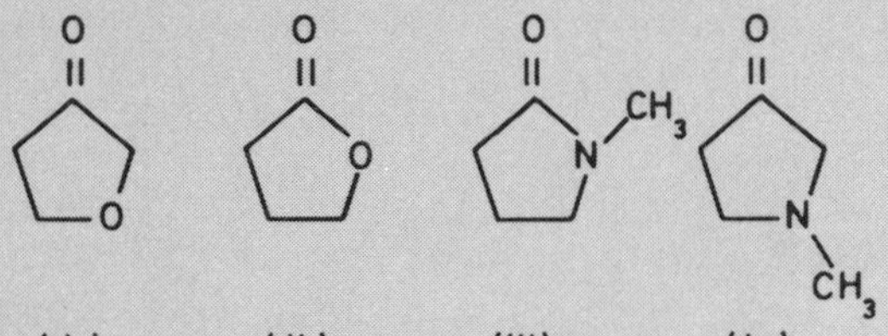

(i)   (ii)   (iii)   (iv)

Response

Our four structures are 5 membered rings. Did you make use of the
table of carbonyl stretching frequencies in Fig. 7.2a (and the Appendix)? You should also have remembered that the C=O stretching frequency of cyclopentanone is 1750 cm^{-1} which represents an
increase of about 30 cm^{-1} compared with larger ring and linear
compounds.

I would expect structures (i) and (iv) to be at approximately the
same frequency as cyclopentanone and hence we have two values
to choose from: 1750, 1745 cm^{-1}.

The choice is not easy, but since the inductive effect from the more
electronegative oxygen is likely to be greatest at the C=O bond I
would give the oxygen compound (i) the highest value.

I would assign the cyclic ester to 1775 cm^{-1}. Esters usually have
values greater than those for the corresponding ketones. This leaves

the cyclic amide (*iii*) at 1700 cm^{-1}, which is reasonable when you recall that amides usually have values below 1700 cm^{-1}.

My answer is therefore

(*i*) 1750,   (*ii*) 1775,   (*iii*) 1700,   (*iv*) 1745 cm^{-1}.

SAQ 7.3a

Examine the spectrum of nonane in Fig. 7.3a and describe the vibrations corresponding to the absorptions marked A, B and C.

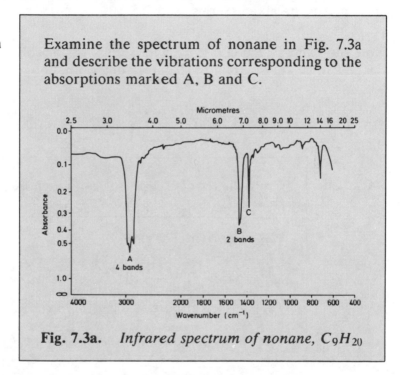

Fig. 7.3a.   *Infrared spectrum of nonane, C_9H_{20}*

Response

The absorption marked A clearly shows four bands, at approximately 2960, 2930, 2870 and 2850 cm^{-1}. Those at 2960 and 2870 are the antisymmetric and symmetric stretching absorptions of the methyl groups, the others are corresponding bands from the methylene groups.

The absorption marked B shows two bands at 1465 and 1450 cm^{-1}. The higher one is the CH_2 deformation, and the lower the antisymmetrical CH_3 deformation.

The band marked C is a singlet and absorbs at 1380 cm^{-1}, this is the symmetrical CH$_3$ deformation.

Note the consistency of these values with those in the main text. They seldom vary much from the quoted values, so check the calibration of your instrument if you get values varying by more than 5 cm^{-1} across the alkane series.

Note the band at 720 cm^{-1}, we will discuss this absorption in the text, it can be very helpful in structure determination.

****

SAQ 7.3b   From the list of compounds below, choose those that

(*i*)   absorb at 720 cm^{-1},

(*ii*) absorb at 1375 cm^{-1},

(*iii*) absorb at both 720 and 1375 cm^{-1},

(*iv*) do not absorb at either 720 or 1375 cm^{-1}.

2-methylbutane

cyclopentane

2-methyloctane

3-methylpentane

butane

1,1-dimethylcyclohexane

Response

2-methylbutane has the structure $(CH_3)_2CH.CH_2CH_3$ and therefore contains a gem-dimethyl group. It does not contain a chain of four CH_2 groups and is therefore in category (*ii*).

Cyclopentane is a five membered ring containing a chain of five CH_2 groups. It will therefore absorb at 720 cm^{-1} and is in category (*i*).

2-methyloctane has the structure $(CH_3)_2CHCH_2CH_2CH_2CH_2CH_2CH_3$, and contains both a gem dimethyl group and a chain of five CH_2s. It is in category (*iii*).

3-methylpentane has the structure $CH_3CH_2CHCH_2CH_3$
$$|$$
$$CH_3$$

and does not contain either feature, so is in category (*iv*).

Butane has the structure $CH_3CH_2CH_2CH_3$ and is in category (*ii*).

1,1-dimethylcyclohexane has the structure below and contains both structural features. It absorbs at both 720 and 1375 cm^{-1} and is, therefore, in category (*iii*).

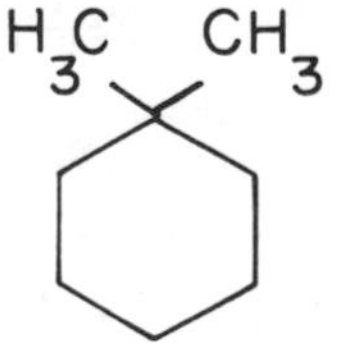

****

SAQ 7.3c   Arrange each of the two sets of compounds below in order of increasing C=C stretching frequency.

(*i*)

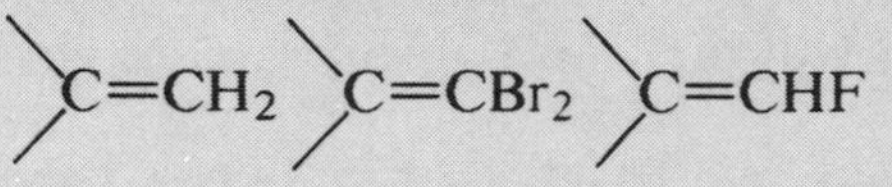

$$\text{\textbackslash}C=CH_2 \quad \text{\textbackslash}C=CBr_2 \quad \text{\textbackslash}C=CHF$$

(a)   (b)   (c)

(*ii*)

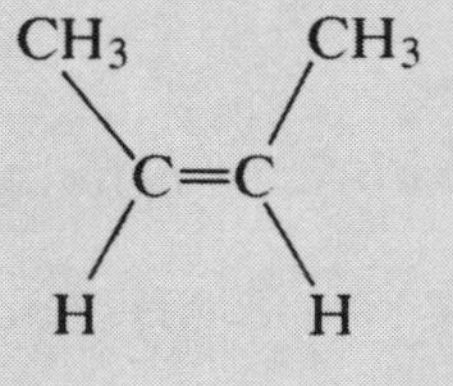

$$\begin{array}{cc} CH_3 & CH_3 \\ \diagdown & \diagup \\ C & = C \\ \diagup & \diagdown \\ H & H \end{array}$$ cyclopentene cyclobutene

(a)   (b)   (c)

Response

(*i*)

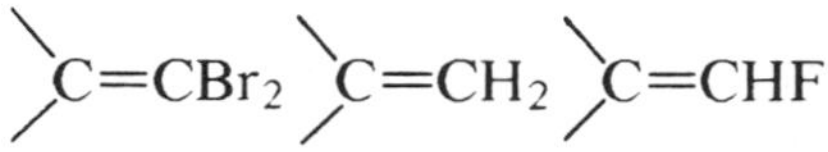

$$\text{\textbackslash}C=CBr_2 \quad \text{\textbackslash}C=CH_2 \quad \text{\textbackslash}C=CHF$$

Fluorine increases the C=C stretching frequency while
bromine decreases it compared to hydrogen, so the order above
would be expected.

(*ii*) cyclobutene then cyclopentene then

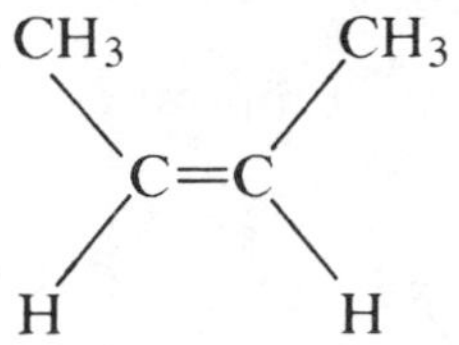

$$CH_3 \qquad CH_3$$
$$\diagdown \qquad \diagup$$
$$C = C$$
$$\diagup \qquad \diagdown$$
$$H \qquad\qquad H$$

cis-but-2-ene.

Remember that cyclohexene absorbs at the same frequency as open-chain alkenes but as the ring size decreases the frequency also decreases (cyclopropenes behave abnormally).

****

SAQ 7.3d

Which of the pairs of alkenes given below could you distinguish using the spectral information provided by out-of-plane bending vibrations of alkenes?

(*i*)   *cis*-But-2-ene and But-1-ene,

(*ii*) *trans*-Hex-2-ene and Cyclohexene

(*iii*) 2,3-Dimethyl-but-2-ene and 2-Methyl-but-2-ene

(*iv*) Propene and 2-Methylpropene

Response

(i) The structures of the compounds are as follows:

cis But-2-ene   But-1-ene

This should be very easy. The first compound should have one absorption between 700 and 800 cm^{-1}, while the second should have three bands, one between 930 and 1000 cm^{-1}, one between 880 and 930 cm^{-1} and one between 600 and 700 cm^{-1}.

(ii) Cyclohexene can be treated as a *cis*-alkene, therefore it should absorb as but-1-ene, ie have three bands. *Trans*-Hex-2-ene has the structure:

and should have one band between 930 and 1000 cm^{-1}

(*iii*) The structures are as follows:

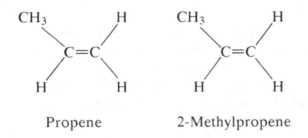

2,3-Dimethyl-but-ene 2-Methyl-but-2-ene

The first of these should not absorb in this region, while the second should give one band between 820 and 880 cm^{-1}.

(*iv*) The structures are as follows:

Propene 2-Methylpropene

Propene will give four bands and 2-methylpropene only two. These deformation bands are therefore very useful!

****

SAQ 7.3e   Which of the following molecules would be expected to have the strongest C≡C stretch?

(*i*)   Ph—C≡C—H

(*ii*) CH_3—C≡C—CH_3

(*iii*) CH_3CH_2—C≡C—CH_3

Which would be weakest?

Response

The intensity of absorption depends on the change in dipole moment during the stretch. In this case the most polar molecule on the list is phenylethyne, (*i*), which will therefore have the most intense band. The absorption will be absent in but-2-yne because of symmetry and will be very weak in pent-2-yne (*iii*).

****

SAQ 7.3f Examine the three spectra in Figs. 7.3f–7.3h. They are isomeric disubstituted benzenes. Which is 1,2 which 1,3 and which 1,4?

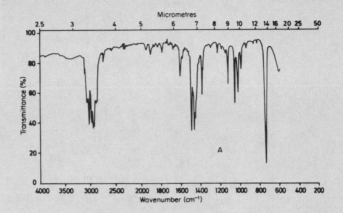

Fig. 7.3f. *Infrared spectrum of a disubstituted benzene: Compound A*

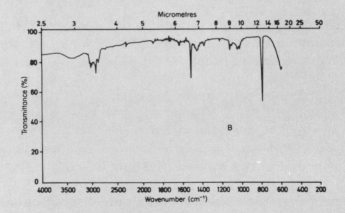

Fig. 7.3g. *Infrared spectrum of a disubstituted benzene: Compound B*

→

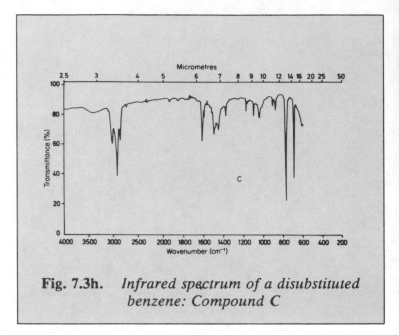

Fig. 7.3h.   *Infrared spectrum of a disubstituted
benzene: Compound C*

Response

These compounds are very easily differentiated using the C—H deformation bands between 600 and 1000 wavenumbers. On examination of the spectra I found,

Compound A: one band at 734 cm^{-1}

Compound B: one band at 787 cm^{-1}

Compound C: two bands at 680 and 760 cm^{-1}

Use of Fig. 7.3e tells us that the only disubstitution pattern giving
two peaks in this region is 1,3. It also tells us that 1,2 and 1,4 give
one peak with 1,2 absorbing at lower frequency than 1,4.

Hence A is 1,2-disubstituted, B is 1,4-disubstituted and C is 1,3.

All three spectra were run as liquid films – note the evaporation of Compound B.

The spectra are of 1,2-, 1,3-, and 1,4-dimethylbenzene.

****

SAQ 7.3g

> Given that a molecule contains oxygen, how would you use an infrared spectrum to find out whether the compound was an acid, phenol, alcohol or ether?

Response

Ethers can easily be identified because they are the only compounds in our list that do not absorb above 3200 cm^{-1}.

Carboxylic acids have very broad absorption from the O—H stretch between 2500 and 3500 cm^{-1}. They are readily distinguished from alcohols and phenols where OH absorption bands, though normally desccribed as broad, are sharp relative to those of carboxylic acids.

Both phenols and alcohols absorb in the OH stretching region and can only be differentiated by the benzene ring absorption of the phenol. If the alcohol is an aromatic alcohol we obviously need a new approach and a simple chemical test is often the answer. Try sodium carbonate!

****

SAQ 7.3h   Do you consider it possible to differentiate the three classes of amine (primary, secondary and tertiary) using N—H stretching absorption?

Response

The short answer is yes. Primary amines contain two N—H bonds and therefore will have symmetric and anti-symmetric stretching frequencies, ie two bands. They occur about 100 cm^{-1} apart. In secondary amines only one N—H bond is present so a simple one-peak absorption is observed. Tertiary amines will not absorb in this region since no N—H bonds are present.

****

SAQ 7.3i   Look back at the correlation chart for the C=O stretching frequency, Fig. 7.2a. What other classes of compound absorb in the same range as saturated aldehydes and ketones? Is it possible to distinguish these compounds using other absorptions likely to be present in the spectrum?

Response

For saturated aldehydes and ketones other classes of compounds absorbing in the same region are carboxylic acids and two acid derivatives – esters and anhydrides. Acids contain the OH group, which, as you saw in Part Five is strongly hydrogen bonded, giving a very broad and characteristic absorption in the region 2300–3500 cm^{-1}. There should therefore be no confusion. Acid anhydrides have two bands in the carbonyl region and are therefore unlikely to cause trouble either. Esters are a little more difficult but contain the

C—O bond, which you saw could be assigned in alcohols and phenols, so perhaps the same is possible here. Go back to the main text, we shall see later if this problem can be solved.

****

SAQ 7.3j

> The 'base' frequency for an open-chain or saturated ketone is 1719 cm^{-1}. The frequency of an aldehyde carbonyl group is found to be higher than this. Explain in a sentence why this is so.

Response

Saturated ketones have two CH_2 or CH_3 groups attached to the carbonyl carbon. Aldehydes have only one of these. The CH_2 or CH_3 group exerts a +I effect, thus lengthening the C=O bond. This means the bond order is decreased, and the frequency correspondingly decreased also. This effect will be larger in ketones than in aldehydes, therefore ketones should absorb at lower frequency than aldehydes, which is found to be the case.

****

SAQ 7.3k

> Examine the spectrum of 2-methoxybenzaldehyde in Fig. 7.3i. Can you see that the band for the aldehyde C—H str is a doublet, even though there is overlap of the high frequency peak with the alkane C—H str? I hope so.
>
> Describe the effect responsible for the doublet formation.

Response

Fermi resonance is responsible. This is the coupling of a funda-mental with an overtone band, the two vibrations being in the same place.

The doublet is centred on about 2780 cm^{-1}, so we are looking for a band around 1400 cm^{-1}. There is a doublet in this region with peaks at 1382 and 1386 cm^{-1}. One of these is a CH$_3$ deformation band – perhaps the lower of

the two. The higher is the $-\overset{\displaystyle O}{\underset{\displaystyle H}{C}}$ in plane band.

The latter must be responsible since the CH$_3$ deformation is not in the same plane as the aldehyde C—H str. Twice 1386 is 2772 cm^{-1} which is close to the centre of our doublet.

****

SAQ 7.31

> Monomeric carboxylic acids show C=O stretch at 1760 cm^{-1}, while in the solid state this ab-sorption occurs at 1710 cm^{-1}. Can you explain this in terms of hydrogen bonding?

Response

In the solid state these compounds exist as hydrogen-bonded dimers with one possible structure (with the vertical lines representing the hydrogen bond) being,

$$
\begin{array}{ccc}
 & \text{O}\text{\tiny||||||} \quad \text{H}-\text{O} & \\
\text{R}-\text{C} & & \text{C}-\text{R} \\
 & \text{O}-\text{H}\text{\tiny||||||} \quad \text{O} &
\end{array}
$$

This leads to symmetric and antisymmetric vibrations. The latter is infrared active and absorbs near 1700 cm^{-1}.

****

SAQ 7.3m   The carboxylate anion can be written,

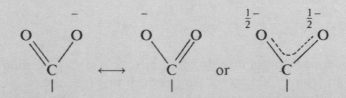

The C—O bonds are therefore equivalent and have the same bond order, intermediate between double and single bonds. Which of the following would you expect to be correct for the C—O absorption?

(*i*) Very intense single peak at 1500 cm^{-1}.

(*ii*) Very intense single peak at 1800 cm^{-1}.

(*iii*) Two bands, one at 1600 and one around 1400 cm^{-1}.

Response

This is analogous to the symmetrical and antisymmetrical modes of
the CH_2 group, two bands would be expected and are found.

****

SAQ 7.3n   Arrange the following lactones in order of decreasing carbonyl frequency.

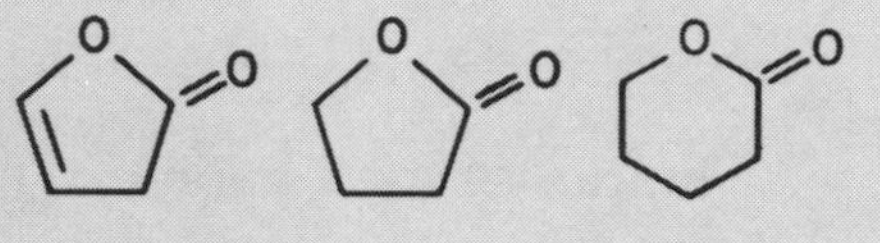

Response

It is found that a double bond next to a singly bonded oxygen atom
in an ester or ketone *raises* the carbonyl frequency. You should also
recall that a six membered ring ketone absorbs at a lower frequency
than a five-membered ring compound. The order is therefore $(a) >
(b) > (c)$.

****

SAQ 7.3o

Look at the structures below and assign each of the following pairs of carbonyl absorptions to one of the compounds, 1865, 1780; 1815, 1750; 1775, 1720 cm^{-1}.

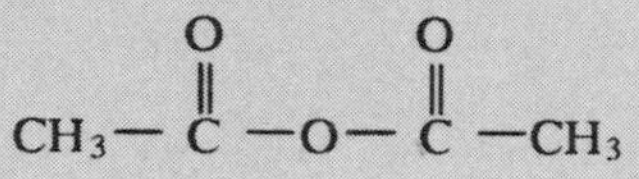

$$CH_3 - \overset{\overset{\displaystyle O}{\|}}{C} - O - \overset{\overset{\displaystyle O}{\|}}{C} - CH_3 \qquad Ph - \overset{\overset{\displaystyle O}{\|}}{C} - O - \overset{\overset{\displaystyle O}{\|}}{C} - Ph$$

(i) 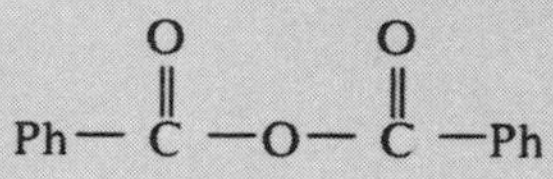   (ii)

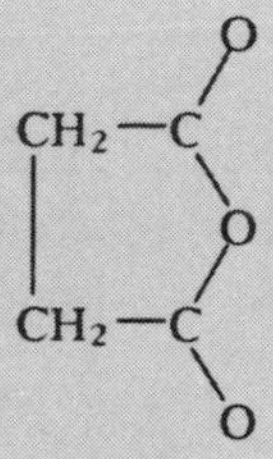

(iii)

Response

You should be becoming fairly familiar with these ideas by now. 5-membered ring compounds absorb at higher frequency than 6-rings or open chain compounds and conjugated aromatic rings will lower the frequency.

So we would expect,

(a) 1815, 1750 cm^{-1}

(b) 1775, 1720 cm^{-1}

(c) 1865, 1780 cm^{-1}

The intensities of the two bands can also be used to distinguish ring compounds from open chain compounds. In small ring compounds it is found that the intensity of the lower band is always much more intense than the band at higher frequency. The opposite is the case for open-chain compounds. The intense bond in ring compounds corresponds to the antisymmetric vibration.

****

SAQ 7.3p

Examine the structures below and choose one of the following carbonyl stretching frequencies for each, 1650, 1660, 1685 cm^{-1}.

The samples were at high dilution in a non-polar solvent.

(*i*) $CH_3.NH.CO.Ph$

(*ii*) $Ph.NH.CO.CH_3$

(*iii*) $CH_3.NH.CO.CH_3$

Response

All the samples were prepared under conditions in which hydrogen bonding can be ignored. All the compounds are secondary amides and compound (*iii*) may be regarded as the standard.

In compound (*i*) the aromatic ring is conjugated to the carbonyl group which results in a lowering of the C=O frequency. Just as for ketones, we would expect that competition for the lone pair electrons on nitrogen by the aromatic ring in (*ii*) would raise the C=O stretching frequency.

The correct assignment of frequencies to the structures is therefore:

(*i*)   1650 cm−1

(*ii*)   1685 cm−1

(*iii*) 1660 cm−1

****

SAQ 7.3q

Match the following compounds with the spectral details below.

$Ph.CH_2.CO.CH_3$

4-Methylbenzaldehyde

$Ph.CH=CH.CO_2H$

$Ph.CH_2.CO.NH_2$

(*i*)   2800, 2680, 1680, 750, 810, no absorption above 3100 cm^{-1}

(*ii*) 3500–2500 (broad), 1680, 1630, 710, 760 cm^{-1}

(*iii*) 3400, 3200, 1640, 690, 740 cm^{-1}

(*iv*) 1710, 680, 740, no absorption above 3100 cm^{-1}

Response

I would first look for absorption above 3100 cm^{-1}, since two of the compounds should absorb there, but in characteristically different ways. The acid should give a broad absorption over a wide range, and the amide should be sharper absorption which may be two peaks, depending on sampling conditions.

Hence the acid, Ph.CH=CH.CO$_2$H corresponds to the spectral details (*ii*). The bands reported at 1680, 1630, 710 and 760 are due to C=O str, C=C str, and two out-of-plane C—H bending vibrations from the aromatic ring, respectively.

The amide, Ph.CH$_2$.CO.NH$_2$, corresponds to (*iii*) with absorption bands at 3400, 3200, 1640, 690, 740 cm^{-1}. The first two correspond to the N—H symmetrical and anti-symmetrical stretching vibrations. The band at 1640 cm^{-1} is the C=O stretch, and the bands at 690 and 740 cm^{-1} the two out-of-plane C—H bending vibrations characteristic of mono-substitution of the ring.

The spectral details in (*i*) list two bands at 2800 and 2680 cm^{-1}, which are characteristic of the C—H str of an aldehyde and we have assigned 4-methylbenzaldehyde.

The ketone, Ph.CH$_2$.CO.CH$_3$, is consistent with the three bands given in (*iv*): 1710, 680, 740, cm^{-1}

****

SAQ 7.3r	Match the following compounds with the N—O stretching frequencies given.
	Nitrobenzene
	4-Nitroaniline
	4-Nitrobenzaldehyde
	(*i*) 1480 and 1319 cm^{-1}
	(*ii*) 1560 and 1360 cm^{-1}
	(*iii*) 1520 and 1355 cm^{-1}

Response

4-Nitroaniline contains the NH_2 group, which is electron donating and should therefore decrease the N—O stretching frequency.

4-Nitrobenzaldehyde contains the —CHO group which is electron attracting and should therefore result in an increase in the N—O stretching frequency.

It is reasonable therefore to assign the frequencies under (*iii*) to nitrobenzene since the other two sets of data show shifts to higher and lower frequencies.

This gives:

Nitrobenzene   :   (*iii*)   1520 and 1355 cm^{-1}
4-Nitroaniline   :   (*ii*)   1560 and 1360 cm^{-1}
4-Nitrobenzaldehyde   :   (*i*)   1480 and 1319 cm^{-1}

****

SAQ 7.3s

> The C—Cl stretching frequency occurs in the range, 600–850 cm^{-1}. Which of the following statements are true?
>
> (*i*)   C—Br stretch occurs between 1000 and 1100 cm^{-1}.
>
> (*ii*) C—I stretch occurs below the normal frequency range.
>
> (*iii*) C—F stretch occurs in the range 1000–1400 cm^{-1}.
>
> (*iv*) Absorption due to C—F stretch are more intense than those due to C—I stretch.

Response

(*i*)   This cannot be correct. You are told that C—Cl stretch absorbs between 600 and 850 cm^{-1}, the substitution of the more massive bromine atom will *reduce* the absorption frequency, usually to less than 600 cm^{-1}.

(*ii*) This must, therefore, be true.

(*iii*) True

(*iv*) The intensity depends on the electronegativity difference between carbon and the halogen. This is greater for C—F than for C—I and this statement is therefore also true.

****

SAQ 7.4a

Oct-1-ene gives the following significant bands in its infrared spectrum. Only the band at 720 cm^{-1} is of weak intensity. (You can check the spectrum in Fig. 7.4c.)

3100 cm^{-1}	C—H str (C=CH$_2$)
2970–2860 cm^{-1}	CH str (alkane)
1650 cm^{-1}	C=C str
998, 915 cm^{-1}	C—H out of plane bending (—CH=CH$_2$)
720 cm^{-1}	—(CH$_2$)$_n$, where $n > 4$

A spectrum of an isomer of oct-1-ene is given in Fig. 7.4d.

Summarise the main differences between the infrared spectra of the two isomers and draw what conclusions you can about the structure of the unknown isomers.

Before you start you will need a few hints.

The C—H stretching frequency for an alkene varies in the range 3000 to 3100 cm^{-1}.

The band at 840 cm^{-1} has an overtone.

The 'split peaks' just below 1400 cm^{-1} are typical of a tertiary butyl group.

Response

— An alkene C—H str is not clearly visible and there is no peak at 3100 cm^{-1}.

— The C=C str is visible at about 1660 cm^{-1} – there is some overlap with the overtone of the 840 cm^{-1} band.

— The 998 and 915 cm^{-1} out-of-plane bands are absent, but there is one reasonably strong absorption in this region at 840 cm^{-1}.

— The 720 cm^{-1} band is absent.

What do we know already?

— The compound is an alkene, C_8H_{16} and it contains a tertiary butyl group.

What does the spectrum tell us?

— The evidence for an alkene hydrogen comes solely from the possible out-of-plane bending vibration at 840 cm^{-1}.

— The absent cm^{-1} band confirms the presence of considerable branching such as for the tertiary butyl group.

We have therefore $\diagdown$C=CH— and $(CH_3)_3C$— leaving us to account for C_2H_6.

This suggests either

$$(CH_3)_3C-\underset{\underset{CH_3}{|}}{C}=C\diagup^{H}_{\diagdown CH_3} \qquad or \qquad (CH_3)_3C-CH=C\diagup^{CH_3}_{\diagdown CH_3}$$

$$(a) \qquad\qquad\qquad\qquad (b)$$

If you got this far you did very well. The answer is in fact 2,4,4-trimethylpent-3-ene (structure *b*).

****

SAQ 7.4b   Look at the spectrum in Fig. 7.4f.

(*i*) Write down the frequency of the strongest band and the group you consider to be responsible for it.

(*ii*) There are many types of this group in organic molecules. What types can you rule out from the frequency exhibited here?

Response

(*i*) The strongest band is at 1705 cm^{-1} and results from the stretching mode of a carbonyl group.

(*ii*) This frequency value rules out all types of carbonyl compound other than ketones, aldehydes, carboxylic acids or esters.

****

SAQ 7.4c   Refer to Fig. 7.4f again and remember we have a ketone.

(*i*)   Is this a saturated or unsaturated molecule?

(*ii*) The C—H stretching absorption in this molecule is a doublet. Is this because,

(*a*) one peak is from the C—H stretching mode of the CH_3 group and the other from the CH_2 group?

(*b*) the C—H stretching frequency is split by Fermi resonance with a band at 1400 cm^{-1}?

(*c*) on a low resolution instrument the symmetrical and anti-symmetrical frequencies of the CH_2 and CH_3 groups are superimposed?

(*iii*) Is this a cyclic or open-chain molecule?

Response

(*i*)   There are no absorption bands just above 3000 cm^{-1} in the C —H str region, so this is a saturated molecule.

(*ii*) This band is always a doublet in saturated molecules. The third option is correct. A high resolution instrument would resolve this band into two peaks each from CH_2 and the CH_3 groups.

(*iii*) Difficult to say, but it is not a four or five membered ketone. The strong band at 1380 cm^{-1} indicates a high percentage of CH_3 groups compared to CH_2 groups, so simple ring
compounds containing only CH_2 groups are also ruled out.
It is either highly branched or a very simple molecule like
$CH_3COCH_2CH_3$.

****

SAQ 7.4d   Calculate the number of DBEs in the following
molecules,

(*i*)   Benzene

(*ii*) C_4H_6

(*iii*) $C_7H_{13}NO$

(*iv*) C_4H_3BrO

Response

I trust you used the formula

$$DBE = 1 + \frac{(2N_c - N_h - N_x + N_n)}{2}$$

(*i*)   For benzene C_6H_6, we have $N_c = N_h = 6$, and N_x and N_n
= 0, so

$$DBE = 1 + \frac{(12 - 6)}{2} = 4$$

This is correct: 3 for unsaturation, 1 for the ring.

(*ii*)   In this case $N_c = 4$ and $N_h = 6$ so the DBE = 2. This is obvious by inspection since the fully saturated analogue would be C_4H_{10}.

(*iii*) $C_7H_{13}NO$

$N_c = 7$, $N_h = 13$, $N_n = 1$, and oxygen does not count.

Hence DBE = 2

(*iv*)   DBE = 3

****

SAQ 7.4e   Interpret the major bands in the spectrum of this compound (C_7H_9N) (Fig. 7.4g), run as a liquid film. If possible suggest a structure or a list of possible structures consistent with this spectrum.

Response

Following my advice from Section 7.1 I would first look for major bands at wavenumbers greater than 1500 cm^{-1}. We see an absorption at 1520, one at 1610 and another at 3480 cm^{-1}, and of course those from C—H stretching (aliphatic and aromatic). The sharp band at highest frequency could be O—H stretch, but since you were told that this is a liquid film spectrum the O—H absorption would be a broad band because of hydrogen bonding. It must therefore be N—H stretch. C_7H_9N has no oxygen anyway! We have found the functional group for nitrogen already. This is an obvious strategy, always find the functional groups given the formula. We therefore have an amine. The multiplicity of this band means that it is almost certainly a secondary amine.

The band at 1610 cm^{-1} could be N—H bend but is more likely to be the ring breathing frequency of the benzene ring, or both. The 1520 cm^{-1} absorption is an aliphatic deformation band, this is confirmed by the doublet below 3000 cm^{-1}.

The low frequency end of the spectrum tells us we have a mono-substituted benzene ring.

The low frequency end of the spectrum tells us we have a mono-substituted benzene ring.

The information we have so far then tells us we have:

C_6H_5— a mono-substituted benzene, and a —NH— group.

If we look at the formula, C_7H_9N, we have already accounted for C_6H_6N, leaving CH_3. There is only one possibility, N-methylaniline.

****

SAQ 7.4f   Fig. 7.4k is the infrared spectrum of a liquid, of formula C_6H_6NCl.

(*i*)   How many DBEs are present, what does this suggest about the structure?

(*ii*) Is there evidence for a benzene ring in the molecule?

(*iii*) What functional group(s) are present?

Response

(*i*)   There are four double bond equivalents.

(*ii*) This suggests a benzene ring. I presume the spectrum told you this too.

(*iii*) There is nitrogen present, and this is confirmed as a primary amine by the peaks above 3000 cm^{-1} and the N—H bend at 1625 cm^{-1}. The chlorine is also attached to the benzene ring.

****

EXAMPLES FOR FURTHER PRACTICE

My interpretation of the ten Spectra are given below. I'm sure its more detailed than yours, but then I've had a lot more practice. I do hope, however, that you were able to make a decent effort at some of them.

Practice Example, a

Given

An infrared spectrum of an unknown liquid [Bp 108 °C, Molecular Formula C$_4$H$_{10}$O.]

Response

This is a saturated molecule. There are no C—H stretching absorptions above 3000 cm^{-1}.

The broad peak between 3200 and 3700 cm^{-1} tells me this is an alcohol or phenol. A C$_4$ alcohol fits the molecular formula.

Is this straight chain or branched? There is no peak at 720 cm^{-1}, so it must be branched. There is also a doublet at 1375 and 1385, indicating a gem-dimethyl or isopropyl group. The very strong band at 1040 cm^{-1} is a C—O stretch.

The compound must be 2-methylpropan-1-ol, $(CH_3)_2CHCH_2OH$.

Practice Example, b

Given

An infrared spectrum of an unknown liquid, recorded as a 5% solution in CCl_4. [Bp 122 °C, Molecular Formula C_8H_{16}.]

Response

Since this is a CCl_4 solution spectrum we can ignore the negative peak between 720 and 820 cm^{-1}, due to imbalance between the beams of the spectrometer.

There is one double bond equivalent in the structure, it cannot therefore contain a benzene ring. There is a peak above 3000 cm^{-1}, however, so this must be an alkene. There is a weak C=C stretching absorption at 1660 cm^{-1} to confirm this. There is no real sign of chain branching in the 1400–1300 cm^{-1} region.

There are two strong C—H deformation bands at 700 and 965 cm^{-1}. There is also a shoulder on the 700 band at about 720 cm^{-1} indicating $(CH_2)_4$.

This does not contain the CH=CH$_2$ group, it must be a mixture of cis and trans disubstituted alkenes. There is only one possibility that fits all these facts, *cis* and *trans* oct-2-ene, $CH_3.(CH_2)_4.CH=CH.CH_3$.

Practice Example, c

Given

Two infrared spectra of an unknown liquid recorded as (*i*) a 5% and (*ii*) a 1% solution in CCl_4. [Bp 202 °C, Molecular Formula $C_6H_{12}O_2$.]

Response

One double bond equivalent. The broad absorption at 2300–3600 cm^{-1} tells me that this is a carboxylic acid. The $C=O$ group accounts for the single DBE. This also accounts for the dilution effects. the appearance of the sharp peak at 3550 cm^{-1} and the splitting of the $C=O$ stretching absorption. This is the appearance of monomer at higher dilution.

The peak at 720 cm^{-1} tells us we have at least four CH_2 groups in a straight chain. There seems to be no evidence of chain branching. The molecule is hexanoic acid.

Practice Example, d

> Given

> An infrared spectrum of an unknown liquid recorded as a liquid film. [Molecular Formula $C_6H_{12}O$.]

Response

I did the calculation and found one double bond equivalent. There is a carbonyl stretching absorption at 1730 cm^{-1} and aliphatic $C—H$ stretching absorptions below 3000 cm^{-1}. There is no peak at 720 cm^{-1} so we do not have a long-chain.

It is not an acid, so must be a ketone or aldehyde. There are no $C—H$ stretching absorptions for an aldehyde. This is an aliphatic ketone. We cannot get much more information. It is in fact methyl butyl ketone.

Practice Example, e

Given

An infrared spectrum of an unknown pale yellow liquid [Bp 210 °C, Relative Molecular Mass 123.]

Response

The molecule is fairly obviously aromatic. The C—H stretching frequencies are above 3000 cm^{-1}. The peaks just below 3000 cm^{-1} are probably overtones from the very strong bands at 1350 and 1530. The peaks at 700 and 850 cm^{-1} tell me this is a mono-substituted benzene, and the two very strong bands at 1350 and 1530 cm^{-1} tell me that a NO_2 group is present. the liquid is nitrobenzene. This fits with the relative molecular mass.

Practice Example, f

Given

An infrared spectrum of an unknown low melting solid recorded as a nujol mull. [Molecular Formula $C_8H_{12}O$.]

Response

This looks quite difficult, let's work out the number of DBEs.

$$DBE = 1 + \frac{(2N_c - N_h - N_x + N_n)}{2}$$

Here $N_c = 8$ and $N_h = 12$, so DBE = 3

Not enough for a benzene ring. These must be a combination of double and triple bonds.

There is no carbonyl group present. There is an O—H stretching band present between 3100 and 3500 cm^{-1} and a monomer band at 3620 cm^{-1}. The sharp peak at 3320 cm^{-1} is unusual and must be significant. There is also a very weak peak at 2330 cm^{-1}. This could well be —C≡CH. We still need another DBE. There are no C=C bonds present. It must be a ring. The compound is in fact,

Practice Example, g

Given

An infrared spectrum of an unknown compound, recorded as a solution in $CHCl_3$. [Molecular Formula C_7H_7NO.]

Response

Strong doublet at 3180 and 3380 cm^{-1} suggests an NH_2 group. There is also a C=O stretching absorption at 1665 cm^{-1}. It looks like a primary amide, ie a compound containing the $CONH_2$ group. It also contains a benzene ring. The molecule is actually benzamide.

Practice Example, h

Given

An infrared spectrum of an unknown compound, recorded as a liquid film.

Response

Another doublet at 3300 and 3380 cm^{-1}, very similar to the previous spectrum, but note the change in intensity. This time there is no carbonyl group. This therefore is a primary amine. It looks aliphatic, the peaks at 1380 and 1470 cm^{-1} confirm this. There is no sign of chain branching. The molecule is hexylamine.

Practice Example, i

Given

An infrared spectrum of a compound containing chlorine.

Can you explain the doublet at 1740, 1785 cm^{-1}.

Response

This is an aromatic molecule, the doublet at 1600 cm^{-1} and the C—H deformation bands around 650 and 880 cm^{-1} point to a mono-substituted benzene. There are no C—H stretching absorptions below 3000 cm^{-1}, so no alkyl groups are present.

There is a doublet carbonyl absorption at 1740 and 1785 cm^{-1}. This must be an acid chloride. It is in fact benzoyl chloride.

The doublet structure of the C=O stretching frequency is very common in aromatic acid chlorides. This is probably Fermi resonance with the overtone of the band at 880 cm^{-1}.

Practice Example, j

Given

An infrared spectrum of an unknown liquid. [Bp 229 °C. Molecular Formula $C_{10}H_{16}O$.]

Response

A fairly complex molecule, so we are not going to get a complete structure.

First let's calculate DBEs.

$$DBE = 1 + \frac{(2N_c - N_h - N_x + N_n)}{2}$$

Here $N_c = 10$ and $N_h = 16$, this gives DBE $= 3$

There is no triple bond present.

There is evidence of unsaturation both in the C—H stretching region and in the double bond region. There are three peaks between 1600 and 1700 cm^{-1}. There could also be an aldehyde C—H stretching frequency present, somewhat obscured by other C—H stretching frequencies.

The band at 1680 cm^{-1} must be a carbonyl group and must be conjugated. Similarly there seems to be both non-conjugated and conjugated C=C double bonds present. These absorb at 1640 and 1620 cm^{-1} respectively.

The C—H deformation region tells me that the double bonds are tri-substituted. I don't think you can get much more information. The compound is citral, a naturally occurring terpene of formula,

$$\begin{array}{c} CH_3 \\ \diagdown \\ C=CH.CH_2.CH_2.\overset{\displaystyle CH_3}{\overset{|}{C}}=CH.CHO \\ \diagup \\ CH_3 \end{array}$$

Units of Measurement

For historic reasons a number of different units of measurement have evolved to express quantity of the same thing. In the 1960s, many international scientific bodies recommended the standardisation of names and symbols and the adoption universally of a coherent set of units—the SI units (Système Internationale d'Unités)—based on the definition of five basic units: metre (m); kilogram (kg); second (s); ampere (A); mole (mol); and candela (cd).

The earlier literature references and some of the older text books, naturally use the older units. Even now many practicing scientists have not adopted the SI unit as their working unit. It is therefore necessary to know of the older units and be able to interconvert with SI units.

In this series of texts SI units are used as standard practice. However in areas of activity where their use has not become general practice, eg biologically based laboratories, the earlier defined units are used. This is explained in the study guide to each unit.

Table 1 shows some symbols and abbreviations commonly used in analytical chemistry. Table 2 shows some of the alternative methods for expressing the values of physical quantities and the relationship to the value in SI units.

More details and definition of other units may be found in the *Manual of Symbols and Terminology for Physicochemical Quantities and Units*, Whiffen, 1979, Pergamon Press.

Table 1   *Symbols and Abbreviations Commonly used in Analytical Chemistry*

Å	Angstrom
$A_r(X)$	relative atomic mass of X
A	ampere
E or U	energy
G	Gibbs free energy (function)
H	enthalpy
J	joule
K	kelvin ($273.15 + t\,°C$)
K	equilibrium constant (with subscripts p, c, therm etc.)
K_a, K_b	acid and base ionisation constants
$M_r(X)$	relative molecular mass of X
N	newton (SI unit of force)
P	total pressure
s	standard deviation
T	temperature/K
V	volume
V	volt ($J\ A^{-1}\ s^{-1}$)
$a, a(A)$	activity, activity of A
c	concentration/ mol dm^{-3}
e	electron
g	gramme
i	current
s	second
t	temperature / °C
bp	boiling point
fp	freezing point
mp	melting point
$\approx$	approximately equal to
$<$	less than
$>$	greater than
$e, \exp(x)$	exponential of x
$\ln x$	natural logarithm of x; $\ln x = 2.303 \log x$
$\log x$	common logarithm of x to base 10

Table 2 *Alternative Methods of Expressing Various Physical Quantities*

1. **Mass (SI unit : kg)**

$$g = 10^{-3} \text{ kg}$$
$$mg = 10^{-3} \text{ g} = 10^{-6} \text{ kg}$$
$$\mu g = 10^{-6} \text{ g} = 10^{-9} \text{ kg}$$

2. **Length (SI unit : m)**

$$cm = 10^{-2} \text{ m}$$
$$\text{Å} = 10^{-10} \text{ m}$$
$$nm = 10^{-9} \text{ m} = 10\text{Å}$$
$$pm = 10^{-12} \text{ m} = 10^{-2} \text{ Å}$$

3. **Volume (SI unit : m^3)**

$$l = dm^3 = 10^{-3} \text{ m}^3$$
$$ml = cm^3 = 10^{-6} \text{ m}^3$$
$$\mu l = 10^{-3} \text{ cm}^3$$

4. **Concentration (SI units : mol m^{-3})**

$$M = \text{mol } l^{-1} = \text{mol dm}^{-3} = 10^3 \text{ mol m}^{-3}$$
$$\text{mg } l^{-1} = \mu g \text{ cm}^{-3} = \text{ppm} = 10^{-3} \text{ g dm}^{-3}$$
$$\mu g \text{ g}^{-1} = \text{ppm} = 10^{-6} \text{ g g}^{-1}$$
$$\text{ng cm}^{-3} = 10^{-6} \text{ g dm}^{-3}$$
$$\text{ng dm}^{-3} = \text{pg cm}^{-3}$$
$$\text{pg g}^{-1} = \text{ppb} = 10^{-12} \text{ g g}^{-1}$$
$$\text{mg\%} = 10^{-2} \text{ g dm}^{-3}$$
$$\mu g\% = 10^{-5} \text{ g dm}^{-3}$$

5. **Pressure (SI unit : N m^{-2} = kg m^{-1} s^{-2})**

$$\text{Pa} = \text{Nm}^{-2}$$
$$\text{atmos} = 101\ 325 \text{ N m}^{-2}$$
$$\text{bar} = 10^5 \text{ N m}^{-2}$$
$$\text{torr} = \text{mmHg} = 133.322 \text{ N m}^{-2}$$

6. **Energy (SI unit : J = kg m^2 s^{-2})**

$$\text{cal} = 4.184 \text{ J}$$
$$\text{erg} = 10^{-7} \text{ J}$$
$$\text{eV} = 1.602 \times 10^{-19} \text{ J}$$

Table 3   *Prefixes for SI Units*

Fraction	Prefix	Symbol
10^{-1}	deci	d
10^{-2}	centi	c
10^{-3}	milli	m
10^{-6}	micro	μ
10^{-9}	nano	n
10^{-12}	pico	p
10^{-15}	femto	f
10^{-18}	atto	a

Multiple	Prefix	Symbol
10	deka	da
10^{2}	hecto	h
10^{3}	kilo	k
10^{6}	mega	M
10^{9}	giga	G
10^{12}	tera	T
10^{15}	peta	P
10^{18}	exa	E

Table 4   *Recommended Values of Physical Constants*

Physical constant	Symbol	Value
acceleration due to gravity	g	9.81 m s^{-2}
Avogadro constant	N_A	$6.022\ 05 \times 10^{23} \text{ mol}^{-1}$
Boltzmann constant	k	$1.380\ 66 \times 10^{-23} \text{ J K}^{-1}$
charge to mass ratio	e/m	$1.758\ 796 \times 10^{11} \text{ C kg}^{-1}$
electronic charge	e	$1.602\ 19 \times 10^{-19} \text{ C}$
Faraday constant	F	$9.648\ 46 \times 10^{4} \text{ C mol}^{-1}$
gas constant	R	$8.314 \text{ J K}^{-1} \text{ mol}^{-1}$
'ice-point' temperature	T_{ice}	$273.150 \text{ K exactly}$
molar volume of ideal gas (stp)	V_m	$2.241\ 38 \times 10^{-2} \text{ m}^3 \text{ mol}^{-1}$
permittivity of a vacuum	ϵ_0	$8.854\ 188 \times 10^{-12} \text{ kg}^{-1} \text{ m}^{-3} \text{ s}^4 \text{ A}^2 \text{ (F m}^{-1})$
Planck constant	h	$6.626\ 2 \times 10^{-34} \text{ J s}$
standard atmosphere pressure	p	$101\ 325 \text{ N m}^{-2} \text{ exactly}$
atomic mass unit	m_u	$1.660\ 566 \times 10^{-27} \text{ kg}$
speed of light in a vacuum	c	$2.997\ 925 \times 10^{8} \text{ m s}^{-1}$

Appendix

Four Correlation Tables:

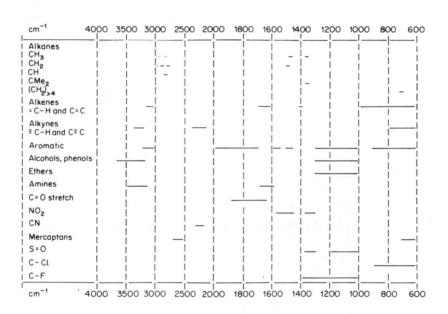

A1 *A Simple Correlation Chart*

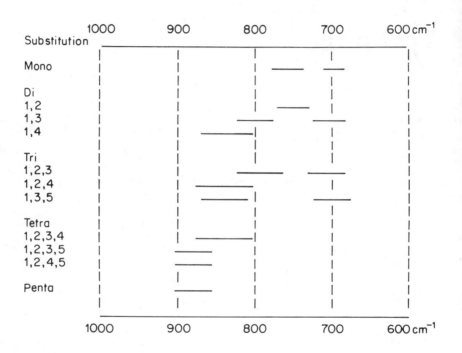

A2   *Out-of-plane Bending Vibrations of Benzenoid Compounds*

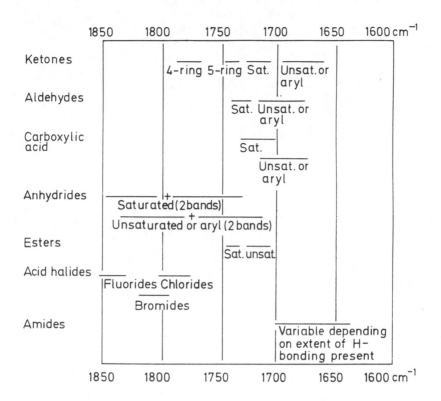

A3 *Carbonyl Stretching Frequencies*

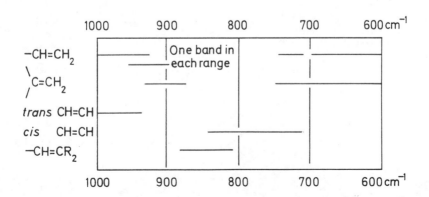

A4 *Out-of-plane Bending Vibrations of Alkenes*